Copyright 2024 by Harvest of Healing, LLC.

Published 2024.

Printed in the United States of America.

All rights reserved.

No portion of this book may be reproduced, stored in a retrieval system, or transmitted in any form or by any means – electronic, mechanical, photocopy, recording, scanning, or other – except for brief quotations in critical reviews or articles, without the prior written permission of the author.

Softcover ISBN 978-1-957077-90-1
Hardcover ISBN 978-1-957077-91-8
Cover image: Shutterstock 2284290583

HARVEST OF HEALING, LLC

Izauh 61™

Publishing assistance by BookCrafters, Parker, Colorado.
www.bookcrafters.net

Owner's Manual of the Universe

The Owner's Manual written by the Universe for the human life must be fully recovered. Lost through the sands of time resulting in disease, decay and death, the ancient language spoken by the Universe is gradually experiencing its long-awaited resurrection. The laws that govern mankind, those that orchestrate health and longevity, are being brought forth by the Master. Now is the time to learn and apply the language all of mankind desperately needs to know.

Izauh 61™

Philippians 3:18-20: *For I have often told you, and now say again with tears, that many live as enemies of the cross of Christ. Their end is destruction; their god is their stomach; their glory is in their shame. They are focused on earthly things, but our citizenship is in heaven, from which we also eagerly wait for a Savior, the Lord Jesus Christ.* (HCS)

~ ~ ~

I Corinthians 6:12-13: *"Everything is permissible for me," but not everything is helpful. "Everything is permissible for me," but I will not be brought under the control of anything. Food for the stomach and the stomach for food," but God will do away with both of them. The body is not for sexual immorality but for the Lord, and the Lord for the body.* (HCS)

I Corinthians 6 tells us the term "sexual immorality" is related to food and the uncontrollable desires related to hunger. That puts a different spin on many verses in Scripture doesn't it! The body is to have the ability to function, to live, and to thrive through a means that comes through Spirit Energy, the Heavenly Gases.

~ ~ ~

I Timothy 5:23-25: *Don't continue drinking only water but use a little wine because of your stomach and your frequent illnesses. Some people's sins are obvious, going before them to judgment, but the sins of others surface later. Likewise, good works are obvious and those that are not obvious cannot remain hidden.* (HCS)

I Timothy is talking about the benefits of concord grape juice. Sins are disturbances in DNA, some are incurred through personal acts and some come through the DNA chain from the acts of ancestors. Either way, consequences will eventually manifest.

In an Attempt to Save the World

EATING YOURSELF TO DEATH
The Effects of Foods
and the Truth Behind Them

HARVEST OF HEALING, LLC

Izauh 61™

INDEX OF CONTENTS

INTRODUCTION..1

CHAPTER I: GOD IS THE GALAXY.........................7

CHAPTER II: HOW DID WE LOSE THE FORCE?............10

CHAPTER III: GOD RESPONDS TO NUMBERS.............12

CHAPTER IV: MIXING IT ALL TOGETHER....................14

CHAPTER V: IS THAT EDIBLE?..16

CHAPTER VI: MUDDY WATER..21

CHAPTER VII: START THE REVOLUTION.....................22

CHAPTER VIII: DO YOU HAVE GAS?...........................24

CHAPTER IX: SETTING THE TABLE................................27

CHAPTER X: WHO SHOULD WE BELIEVE?.................42

CHAPTER XI: BEVERAGES..52

CHAPTER XII: BREADS PASTRIES AND PASTA.............62

CHAPTER XIII: CANDY AND CHOCOLATE.................69

CHAPTER XIV: CONDIMENTS..................................72

CHAPTER XV: DAIRY..73

CHAPTER XVI: FRUIT...77

CHAPTER XVII: GRAINS...80

CHAPTER XVIII: MEATS AND PROTEINS.....................82

CHAPTER XIX: NUTS, SEEDS AND OILS......................93

CHAPTER XX: SNACKS AND MISCELLANEOUS...........95

CHAPTER XXI: SUPPLEMENTS.................................98

CHAPTER XXII: VEGETABLES...................................99

CHAPTER XXIII: APPLIANCES.................................102

CHAPTER XXIV: EMOTIONS....................................104

CONCLUSION..109

INTRODUCTION

Galatians 4:22-26: For it is written that Abraham had two sons, one by a slave and the other by a free woman. But the one by the slave was born according to the impulse of the flesh, while one by the free woman was born as the result of a promise. These things are illustrations for the women represent the two covenants. One is from Mount Sinai and bears children into slavery – this is Hagar. Now Hagar is Mount Sinai in Arabia and corresponds to the present Jerusalem, for she is in slavery with her children. But the Jerusalem above is free and she is our mother. For it is written: Rejoice, childless woman, who does not give birth. Burst into song and shout, you who are not in labor, for the children of the desolate are many, more numerous than those of the woman who has a husband. (HCS)

Every human comes from the root of righteousness (Abraham's seed), yet one group is free from the slavery of sickness and disease and the need for abundance of food by living with the Spirit (Heavenly Gases), and one group is enslaved to the fleshly urges connected to foods (sexual immorality according to I Corinthians) that results in sickness and disease. There is no peace inside the body when slavery to food is involved.

My life seemed to revolve around trying to get my body to conform to the common things in life when it came to eating, sleeping, decisions, clothing styles and much more. After approximately 50 years of the daily struggle of attempting to figure out why my body could not tolerate foods that everyone else seemed to be able to eat with no problem or how so many friends or family members could sleep at any time of day and sleep through any level of commotion, I decided I must settle with the fact that my body was made different than many. Years of allergy testing, energy testing, muscle testing and every other type of testing outside of what Western Medicine provides, I discovered some valuable information. Information that just may explain many of the troublesome issues most people face today. Some of what I share will seem baffling and raise a question of "how does she know this?" but if you allow the information to simmer for a bit, you will begin to see a pattern that has rarely, if ever, been considered.

Many years ago, I learned that everything, literally everything has and is a vibration. With that in mind, I began to consider the vibration of foods when eating them myself and considered how others respond to various foods. Years of collective observation and personal experience has led to some interesting conclusions although I cannot admit to capturing any scientific proof. Subtle patterns seemed to develop, some involved the current moon phase, and some would involve emotions. It all became quite an interesting and admittedly sometimes frustrating quest to discover answers, particularly when I came to realize the elemental gases inside the body were involved. Those are challenging to track when it comes to a response produced in your health.

I was born into a genetic stock that was all too familiar with digestive disturbances. Everything from colicky infants, Hirschsprung's Disease, constipation, diarrhea, intestinal cramping, nausea, colon cancer, stomach cancer and probably everything in between was posted on the genetic bulletin board I worked from most. Milk was the most popular culprit for introducing a tummy or intestinal upset yet the more observation and experience

I had, it became evident there were far more culprits to the all-too-common digestive disturbances. I share many of the discoveries made and most are yet without any explanation other than putting 2 and 2 together. Because of the uniqueness in each person, no two people respond the same to a food yet there seems to be those common threads, and genetics play a big role. I am convinced however, that many common foods are guilty of altering the origins of DNA that has led humanity down the road of disease.

<u>II Chronicles 21:12-15</u>: ...This is what Yahweh, the God of your ancestor David says: "Because you have not walked in the ways of your father Jehoshaphat (one who is judged; govern) *or in the ways of Asa* (man of sorrows; healer) *king of Judah but have walked in the way of the kings of Israel, have caused Judah and the inhabitants of Jerusalem* (peace) *to prostitute themselves like the house of Ahab* (brother; fellow member of a social economic mode) *prostituted itself, and also have killed your brothers, your father's family, who were better than you, Yahweh is now about to strike your people, your sons, your wives and all your possessions with a horrible affliction. You yourself will be struck with many illnesses, including a disease of the intestines, until your intestines come out day after day because of the disease.* (HCS) (descriptions added)

How is it that these verses seem to be forgotten when digestive health issues erupt? It seems a blind eye is being turned and people continue marching to the beat of the same deadly drum. Somewhere back through history the ancestral line stopped following the rules! Messages from ancestors are screaming at us from the grave through the genetic mess they have left behind that sends their numerous heirs to the medical facility. I have waded through the muddy mess left for me for over 50 years and have discovered some valuable information along the way. I just pray the information I have come to know will help the masses as the years progress.

Foods are the number one player in the measure of life or death within a human body. From the eating of the forbidden fruit attached to the tree that grew in the middle of the Garden of Eden to supersize servings and buffet style dinners, humanity has walked themselves into a life of sickness and disease. It is becoming obvious to me that much of the DNA damage seen today originated in the consumption of combined food groups. Here is a good place to share the eating habit of a gentleman I know. Since childhood, this now grown man has eaten one specific serving of food at a time on his plate. For example, if his plate contains meat, potato, green beans and bread, he eats all of the green beans, then pivots his plate and eats all of the potatoes, and so on, not necessarily in that order but you get the idea. Could there be an ancient DNA signal that surfaced in this man that brought about this uncommon eating habit? God speaks in mysterious ways and many times God's messages are overlooked.

<u>Romans 8:5-6</u>: *For those who live according to the flesh think about the things of the flesh, but those who live according to the Spirit, about the things of the Spirit. For the mindset of the flesh is death, but the mindset of the Spirit is life and peace.* (HCS)

Note: Hebrew DNA references the origin of human DNA; no inherited contamination or personal exposure to contrary influences in their blood (sin-free). Blood that has been cleansed through the required protocols and laws set forth in Scripture would be Hebrew blood (righteous from the seed of Abraham). The term Samaritan is used to refer to a person who has inherited contamination, sometimes referred to in Scripture as "curse," and often referred to as "sin." Spirit is a term used in Holy Bible Scripture (although the interpretation of the term Spirit in Scripture has been corrupted) and, used in this book to reference the reaction from or product of elemental gases most often present in the Heavens (cosmos) coming together. When Spirit, a balance in required elemental gases, interacts with matter/flesh (material objects), a reaction will develop. This reaction ushers in the term Spirit filled life. In turn, the collection of Heavenly Gases causes the plasma in the blood to become electrically charged. This electrically charged plasma is the Living Water referred to in Scripture. The Living Water keeps the blood and body clean of harmful particles and intercepts the need for the volume of food and drink commonly consumed today. Star Dust is the term used to reference the genetic codes endued with Heaven Energy (Spirit). An abundance of Star Dust in the genetic codes produces the Christ status and the resulting lifestyle I call Spiritual Royal.

CHAPTER I
GOD IS THE GALAXY

I woke early one morning, around 3:00 a.m. and was not able to fall back asleep easily. I realized I was experiencing one of those situations where I should be alert because this was the time when the Spirit is known to deliver messages. I did not remove myself from my bed, nor turn on any lights but quietly waited for any message that might be delivered. I've had these experiences quite often during my life, so it was nothing new. The need to just be still and wait can bring some amazing insight. It is a natural response to want to fall back asleep to complete that rest the physical body may need from the previous day's activity, but God sometimes has a different way of refreshing our body that does not involve sleep. As I was lying still, not really having any thoughts of my own other than how quickly the clock would move and it would soon be time to rise for the day, the following message in the form of thoughts and mental images began to download. Try to picture in your mind what I describe and the whole of life and existence will be easier to understand.

Is there really a "Void," a space that is full of nothingness? I suggest this invisible pulsing emptiness currently called Void is where the Force(s) that control the positioning of the planets, the wonders of nature and the mysteries held within the galaxy exists and the name of that collective group of Forces is God! That glue-like electro-magnetic charge that holds all planets and things in their places yet orchestrates their every motion, is God. That collective

Force that orchestrates the entire existence of the galaxy also directs the seasons and the Heavenly Gases that keep all things in working order. This orchestrated activity that takes place in the darkness of the galaxy produces vibrations. Comparing this activity to music, music must possess a harmonious flow and complimentary hertz vibration or the music becomes less than pleasing to the ears and forces resulting from those vibrations can become chaotic. A similar situation happens within the physical body when the internal gases are either in their proper measure, or not. You have the harmonious flow, or you have chaos taking place inside the body.

We, as human physical bodies, are modeled after the earth. The planet earth and the human physical body both need the elemental gases that are present in the galaxy for proper function and existence. The galaxy, through a means of the Force (God) provides the vital fuel for both the earth and the physical body. Cloud vapors distribute Heavenly Gases that originate from planets within the galaxy to the earth. The Heavenly Gases arrive in the atmosphere where they contact the physical body. The Heavenly Gases are absorbed by the physical body and are carried through the body by the plasma as it moves through the veins. This act of gases flowing throughout the body brings the electrical activity that produces the health the body requires. When the Heavenly Gases are in the proper proportions within the body, the gases produce the hydration necessary for life. Would this process qualify as a combustion reaction?

Genesis 2:7 states that Adam, being the first human form, was made from the dust (or soil) of the ground. This would put Adam, the physical body, in a category of being "earth-like;" what the earth needs, the physical body will need. We see this in both the earth and Adam needing daylight and nighttime. Basically, what happens to or takes place on/in the earth will be in like-kind for the life of Adam. The physical body will (or should) go through the cyclical season of growth for three months (Summer), season of things falling away for three months (Fall), season of rest, receiving or dormancy for three months (Winter), and season of change or awakening for three months (Spring), in unison with the seasons experienced on earth. This is the cycle of the cells in the body and each season brings something new for a period of three months.

We must look at the earth as a whole, the big picture of its existence. The rotation of the earth is orchestrated by gases in the galaxy just like the rising of the sun and moon are orchestrated by gases in the galaxy. Those Heavenly Gases that keep the necessary galaxy movement going, and planets suspended in space and in check are also necessary for the human body to exist. There is no pollution in the galaxy, the atmosphere is clean and sings a harmonious tune, thanks to the Heavenly Gases. The same action will take place within the human physical body when the proper gases are in place. The same way the soil of the earth absorbs and releases gases through various movement and means within its core, is in like manner within the physical body. This is how the human form captures and lives "in His image" (Genesis 1:27). The entirety of what is described above is "God." God (Force) is in all places, at all times and we must learn to receive, use and protect the Force that is to reside in and amongst us. It is then that the epic dance will begin.

As Obi-Wan-Kenobi of Star Wars would say: "May the Force be with You!"

CHAPTER II
HOW DID WE LOSE THE FORCE?

The world is overflowing with activities, sound, foods and fun that rob our body of the Heavenly Gases necessary for proper function and the interior housekeeping the body requires. From car rides to plane rides, carnival rides and elevator rides, chemical based cleaners and synthetic clothing, these things short circuit the activity of the gases our body needs to function in a healthy manner.

Combining various foods in our diet has caused chemical reactions that rob the Heavenly Gases and, in some cases, has initiated creation of some explosive and dangerous gases all their own. Mixing non-compatible chemicals or gases can have devastating results in a laboratory, let alone inside the body. What we don't see, like what goes on inside the body, often contains the most dangerous situations.

There is no simple way to express the dangerous lifestyle that has come about through years of development, advancements and change, and I have expressed some of those harmful aspects in previous writings. It seems at times that nearly everything needs to come to a screeching halt so a new perspective and set of life rules can be put into place. The entire worldly system seems to be a mess, if you are a person who desires to live healthy through your days here on earth and have any chance of living in the great hereafter. The best approach for a possible change is to educate. Education will bring a new enlightenment that will spark new perspectives and usher in the change needed. A new respect for the Force that comes from the galaxy will begin to take shape. And, to this, I say AMEN!

CHAPTER III
GOD RESPONDS TO NUMBERS

Picture with your mind the spiral helix of DNA and each rung in that helix staircase has an assigned number. Taking a stroll along the history of this numerical staircase, here's what we might find: (Note: the numerical sequences in DNA presented are made up as an example.)

In the beginning before damage to DNA occurred, the strands of DNA, those spiral staircase appearing objects, contained uninterrupted sequences of numbers. These numbered staircase strands reside inside the cells. A numerical sequence might be 3456703 and this staircase strand produces blue eyes. Through generations ancestors lacked access to gluten proteins, and this caused ancestors to eat beans or eggs as their staple protein source. This shift in diet caused the numerical sequence to become altered or damaged and now the numerical sequence is 3 56703, number 4 was damaged or removed. This new sequence of numbers produces brown eyes in the descendants. Brown eyes certainly do not seem to be an issue until the cycle of cell duplication continues through generations. Years pass and generations of ancestors were exposed to toxic chemicals and the numerical sequence of 7317317 that produces a properly functioning immune response becomes 731 317 and this new numerical sequence produces inflammation and allergies in the descendants. Enough generations of damaged DNA can have a huge impact on the body. The body is continually copying the numerical rungs of the staircase with damaged or missing numbers. Now, you may think brown eyes, inflammation and allergies are not bad things, which in the bigger picture they seem minor but what is not seen or recognized is these damaged DNA numerical sequences prevent or hinder the Spirit Energies that deliver the Heavenly Gases from reaching the individual. God, the Force, connects with a person through the numerical sequences encoded in DNA. The messages in the cosmos had to be present before the creation of specific things – the numbers were the seed of creation. Properly coded DNA is referred to in Scripture as "righteous," the seed/DNA of Abraham, who received insight directly from God. God must be able to dial a properly working cell (think blood cells) number to be able to communicate with you. This same concept is used every day when we dial a sequence of numbers on our cell phone to reach a desired individual for communication purposes.

Food and drink win the award for inflicting the most damage to the numerical sequence in the staircase. The lust of the flesh has caused great harm. The body does not need food in the quantities commonly consumed day after day. Yes, there are specific circumstances where food intake will be different than the general guidelines may suggest. These are details that will come forth as individuals move along this journey of change.

CHAPTER IV
MIXING IT ALL TOGETHER

Genesis 25:27-33: When the boys grew up Esau became an expert hunter, an outdoorsman, but Jacob was a quiet man who stayed at home. Isaac loved Esau because he had a taste for wild game but Rebekah loved Jacob. Once Jacob was cooking a stew, Esau came in from the field exhausted. He said to Jacob, "Let me eat some of that red stuff, because I'm exhausted." That is why he was also named Edom. Jacob replied, "First sell me your birthright." Look, said Esau, "I'm about to die, so what good is a birthright to me?" Jacob said, "Swear to me first." So he swore to Jacob and sold his birthright to him. Then Jacob gave bread and lentil stew to Esau; he ate, drank, got up and went away. So Esau despised his birthright. (HCS)

Jacob and Esau are telling us a lot of information in these verses. First, Esau portrays an individual who is busy, someone who is out doing strenuous activity and comes home willing to eat anything that is available. Hunger pangs are overcoming any restraint that should be in place. Esau is also a hunter, giving note of being a meat eater. Jacob is more complacent hanging out at home, tending to the necessary chores of the house, which would have included making the daily batch of bread. Isaac and Rebekah, in these verses, represent the physical nature (Isaac) and spiritual or electrical nature (Rebekah). The physical body has a natural desire for heavy, filling meals in an effort to quiet the stomach, particularly after strenuous activity; the spirit/electrical aspects of the body require that we refrain from eating heavy, multi-ingredient (stew) meals. The quieter or more subdued the activity, the more likely we will avoid heavy, multi-food meals. Heavy meals steal the Spirit life from an individual by interrupting the Heavenly Gases within the body. Esau's birthright was handed over because of his inability to refrain from the combination of foods present. The birthright represents the DNA's ability to produce a life free of debilitating disease and untimely death (Soul death and physical death). Your birthright is to be eternal.

What is stew or most any soup? A collection of various foods, such as meats, beans, vegetables, fruit (if it contains tomato), and so on. Eating a combination of foods at any one given meal interferes with the health of the blood, depletes the Star Dust and will eventually remove the ability to produce the Living Water. Buffet style eating hinders the Spirit life within the body and will cause the Hebrew DNA to be diminished. The reference to the color red indicates this situation being linked to the blood. When multiple foods are eaten, chemical responses take place that damage the electrical charge, or light that is present within the blood/plasma. According to these verses the established Food Pyramid should be thrown out.

CHAPTER V
IS THAT EDIBLE?

In 2020 I had a complete genetic panel run. Two months after submission of a few tubes of saliva, a report of approximately 100 pages was generated, each page containing approximately 30 lines of numerical and alphabetical codes, each linked to an inherited gene. I took some time to investigate what some, maybe 100 or so, of the codes represented. It didn't take long before I was wondering how all this mess of disease-bound codes were birthed! My parents lived a generic life and my upbringing was simple. There were less than a half-dozen trips my parents made outside of the United States, other than Dad's time in the navy prior to their marriage. A trip to the Bahamas many years ago and they may have stepped across the border of Canada or Mexico. My Dad did not like airplane travel so long-distance trips were far and few between. Travel to foreign countries did not seem to fit for carrying the blame for the health disruptions so how did my health end up in such a mess? Every organ in my body seemed to have a death message attached to it. One thing I knew for sure was that God did not create humans to be diseased. What would be the point in that? So where did this all come from? My conclusion was this, the very thing all persons on earth do is eat. There had to be a connection to food.

Estimated prevalence of diabetes in America: Total 38.4 million people of all ages had diabetes (11.6% of the population) in 2021. 38.1 million were adults ages 18 years or older. (NIH.gov).

60-70 million Americans are suffering from gastrointestinal (GI) diseases. (Gastro.org)

It is obvious human focus has been upon incorrect eating habits, some of it being intentional and some not intentional, along with consumption of numerous types and styles of what has been labeled as food. It is questionable whether many of these "consumable products" should even enter the human body. Then again, maybe we have an incorrect idea of what food really is. If these foods do qualify as a permissible product for consumption, then why are so many people diseased? This food focus has altered the origin of DNA, produced damaged DNA strands and has led to genetic imprints that cast a heavy burden on those that are the recipients thereof.

What is the definition of food? "Any nutritious substance that people (or animals) eat or drink." This definition brings a few questions to mind: 1) who decides what qualifies as "nutritious;" and 2) why do "substances" qualify as a food? A substance is a particular kind of matter with uniform properties. This definition must be a contributor to the development of various concoctions that are injected with synthetic "nutrients," given a splash of chemical flavoring, then packaged and placed on grocer's shelves for purchase and consumption! Who should be thanked for this foolishness? Following the trail to where the money leads would be a good start.

An important point to bring up is that when Scripture references food it is at times referring to nourishment, particularly when the word "meat" is used. The origin of the word meat comes from Middle English "mete" meaning nourishment, meal or sustenance; Gothic "mats" meaning food; Proto-Germanic "mati" from PIE "mad-i," from root "mad" or moist, wet; Sanskrit "medas" meaning fat, Old Irish "mat" meaning pig. (etymonline.com). I must insert here that the Old Irish pig meaning brings up some interesting thoughts of when Jesus cast the legions into the pigs. Hum. Do legions (a vast, host or multitude) have a connection to the consumption of pork, such as producing numerous health related issues? That's up for debate but interesting, nonetheless.

Many foods contain a history people are not aware of. Is it knowledge that is lacking, or full exposure to the public of scientific studies and tests that is lacking? It is possible that too much faith rests in those who hold awards and certificates that boast of a qualification allowing them to decide what is edible and what is not. People rely on reports generated for and financially backed by the producer of a particular food, or should I say substance, being tested. It only makes sense that the food tested would boast of the benefits and hide any negative results. Detailed testing might actually reveal the downfalls to the food/substance/product and that is not the road the producer or manufacturer wants to travel. Foods presented in many forms and fashions in grocery stores, restaurants, catered events, bakeries and more hold more influence upon the physical body than what one might expect or is being told. How many of these tests on foods are done in combination with another food or drink? Such as, what happens to chocolate when you eat it with a glass of milk or cup of coffee? There are far too many combinations to be had for the extensive amount of time and money it would cost to perform such tests. The entire landscape of food has been altered in one way or another and much of those alterations have resulted in the disease and death experienced today, and for the past 100+ years, more like eons. Eventually the negative influence will catch up with you, or your descendants.

The question of how humans landed in the place they are with respect to food is no mystery. A meager yield to the yearly staple crop and a person will migrate to eating objects outside of the assigned food source for survival reasons. The mystery that needs solved is how to get humans out of the habitual mess they are now in with respect to the type of food they consume and when the beneficial time of day is to eat those foods. Another hurdle to pass is the response the physical body may have to a drastic change in diet. Like any other big project, adjusting the diet must be done gradually lest you spend a great deal of time near the commode!

The American diet mindset is to just throw everything into the belly and believe it is going to properly process. In some respects, this may be how things work but an ancient secret to the process of eating and digesting is that each item put in the mouth, chewed up and swallowed will leave an impression or essence along its journey. The essence that remains will eventually develop into and manifest as an emotion or form of health disruption. That's not even considering the gaseous environment all that food creates.

The Universe communicates to us through manifestations of the acts humanity is taking. Consuming food in an "anything and everything" fashion has created alterations in the structure of relationships or sexual preferences. Remember the reference to sexual immorality in I Corinthians 6:12-13 presented just prior to the Introduction? Sexual immorality is a reference to improper desire for and consumption of food. This form of food related sexual immorality will manifest in actual physical relationships. If "anything goes" in the diet, then it will manifest into an "anything goes" in lifestyle choices.

CHAPTER VI
MUDDY WATER

Many foods result in the blood plasma being contaminated, like what one might see in a dirty pond or raging muddy river. If you have used an ion footbath unit, a pond scum appearing film begins to develop on the top of the water your feet are positioned in during the ion session. Not an overly welcoming site. I have used an ion footbath unit for over 15 years, and I can attest to the pond scum and to the fact that foods eaten will influence the amount of pond scum that appears in your footbath water. With all that said, when the following Chapters reference a food as having an influence on fluid, water or plasma, this is what it is referencing, the likelihood of a muddy scum developing in response to a food or combination of foods. If you eat foods day after day that cause the development of this pond scum, it becomes challenging for the body to keep up with the internal cleansing process and robs the plasma of the electrical charge (light) it is to carry for the health and healing of the body.

<u>Ezekiel 32:2</u>: *"Son of man, lament for Pharaoh king of Egypt and say to him: You compare yourself to a lion of the nations, but you are like a monster in the seas. You thrash about in your rivers, churn up the waters with your feet, and muddy the rivers."* (HCS)

CHAPTER VII
START THE REVOLUTION

Imagine being on a road trip when at mile marker 2024 you discover you have been traveling in the wrong direction to get to your desired destination. What is the most logical thing to do at this point? Turn around and go back in the direction "from whence you came." The road of travel with respect to food consumption has landed people at a destination riddled with pain, suffering and disease. If the current generation and generations to come do not wish to end up at the same destination of disease their ancestors did, they must turnaround. Just like on a road trip, when a turnaround to head in the opposite direction is done properly, you will pass by the various locations and landmarks you previously passed. These locations and landmarks can be in the forms of physical symptoms, i.e., stomach upset, insomnia, etc. but they must be passed by in order to get to the desired destination, one that is free of disease. This is not an easy road trip to make. Many years of traveling in the incorrect direction has made deep ruts and stubborn mindsets that are challenging to overcome. Many will claim that the current diet is working for them. They will report that they have no desire to change and that they feel fine, all the while regularly consuming three or four prescription drugs to help them manage through their day. Certainly, choice of food is personal. The knowledge that is lacking is in the department of how a current food choice and dietary habit will influence that person's descendants, no matter how old the person is. There are mysterious messages that come through the cells in the body. The more time you spend with a person, the closer you are to them in physical contact, the more of their cell communication you will receive. Cells communicate with cells, just like cell phones connect one to another.

CHAPTER VIII
DO YOU HAVE GAS?

As you move through the following chapters, keep in mind the information shared is not based on nutritional guidelines. My experiences have taught me a little about these <u>Heavenly Gases</u> that seem to be a hidden mystery to most avenues of professional healthcare. I will refer to the gas combination as Heavenly Gases speaking of them collectively as I admit to not having the knowledge of their appropriate healthy proportions. Inside tip, the term Spirit in Scripture is a reference to the Heavenly Gases.

There are eleven identified elemental gases that remain stable at room temperature: Hydrogen (H1), Helium (He2), Neon (Ne10), Argon (Ar18), Krypton (Kr36), Xenon (Xe54), Radon (Rn86), Fluorine (F9), Chlorine (Cl17), Nitrogen (N7) and Oxygen (Ox8) (sciencenotes.org).

When the body becomes hungry could it be an indication that the Heavenly Gases are lacking in one element or another? Is our body designed to be a gas driven machine verses a food driven machine? Have we become programmed to think that our daily fuel comes only from food which has evolved us in a direction away from knowing we need the Heavenly Gases for survival? Maybe a look into specific mealtimes in the following chapter will uncover some mysteries in this area.

Everyone becomes a little breezy from the backside at times but the gases I am referring to are the elemental gases necessary for the proper function of the body. Gases are sensitive and need a specific environment to do their job and are only produced and housed in a specific environment. Too much heat or cold, or swift movement will cause a disruption in the gases, even causing them to dissipate. For example: Does the downward movement of an elevator cause your stomach to drop? Heavenly Gases respond to the sensation of falling and interrupt the stability of, I'll use the element numbers 54 and 38 as an example, resulting in that dropping sensation felt in the region on the stomach. If a person experiences this swift elevator movement repeatedly for "x" number of years, could it cause a manifestation of a disease as a result of the change in gases in the body? I knew a young lady who worked in a high rise building and took an elevator up 30+ stories each day to the floor where she worked. Each day of the week she would ride this elevator up and down. After a relatively short number of years, and I will guess her age was maybe late 30s or early 40s, she developed Lupus. Could there be a connection between the daily elevator rides that disrupt the internal Heavenly Gases and the onset of Lupus? These are the type of mysteries that need solved.

Gases are received into the body through foods and through the air we breathe. The gases are attracted to mineral stones and gemstones, placing good reason on wearing authentic jewelry and tossing out the costume jewels. When the body is exposed to various forms of heat, such as outdoor temperatures reaching uncomfortable degrees, the extreme hot or cold temperature of food or drink being consumed, and spicy hot foods, it will contribute to the evaporation or dissipation of the gases needed by the body for proper health and function.

Heat burns up gas just like the gas burner on a stove and exposing the body to temperatures above a specific range, whether on the inside the body or on the outside of the body, will interfere with the internal Heavenly Gases. Say goodbye to saunas and hot tubs! Need I add that certain forms of exercise are included in the dangerous internal depletion of gases. Could the dissipation of Heavenly Gases due to heat be an answer to teenage boys dropping dead on football fields? Yet another area that may require some investigation.

A similar issue occurs when the body is too cold. Eating frozen or ice chilled food or drink cools the body temperature down to a point the Heavenly Gases are hindered in their function and movement. Gases are carried by vapor or fine mist and when that mist becomes too cold it does not move as freely as it should. This concept casts a frown upon cold swimming pools, cold showers and spending any length of time outdoors on bitter cold days.

It goes without saying that a quieter more environmentally friendly lifestyle will benefit your health. The calmer the daily activity is, the more balanced the Heavenly Gases will be. Another way to put it is, to live in the Spirit, walk in the Spirit and have any union with the Spirit requires the appropriate amount and function of Heavenly Gases. The worldly lifestyle truly has contributed to killing off the Spirit life.

CHAPTER IX
SETTING THE TABLE

Where you eat your meal and how you are dressed when you eat your meal have significant impact on the body. Dining tables are to be made of wood, and they should not rest against a wall. Outdoor picnic tables, cloths spread on the beach or in the grass do not qualify as an energetically acceptable place to eat a meal. Table linens made of cotton or linen are encouraged for tablecloths, table runners, placemats and napkins. No paper products are to be used for eating purposes. Paper and plastic products generate static electricity that interferes with the health of the cells. Place settings are to be of Fine China, nothing plastic and no stoneware. All silverware (stainless steel or silver) should match. No eating from large spoons (tablespoon) unless you desire descendants with large mouths. Mix-matched silverware will alter the facial features in the region of the nose and mouth. No chopsticks. Stemware is to be used at the evening meal. Salt and pepper left on a table during a meal will result in digestive disruptions.

The clothing worn during your day will determine the amount of energy you gain or lose while wearing that clothing. If clothing does this in general throughout a day, what is it doing to the processes of eating and digesting foods? Remember, there are Heavenly Gases at play when we eat and drink. Move back through history and it was common for men to be in suits with a form of necktie, more of a bowtie than what is worn today, and ladies were always in dresses. Proper dinner attire is not missing from any Royal table and many business or formal dinners have etiquette requirements as well. Your home dining experience should be the same.

Genesis 43:32: They served him by himself, his brothers by themselves, and the Egyptians who were eating with him by themselves, because Egyptians could not eat with Hebrews, since that is detestable to them. (CSB)

In most situations outside of a restaurant, when you sit at a table with people, everyone enjoys the same foods. This verse is bringing the attention to the separation of food choices between those classified as Hebrew and those who live according to the world (Egyptian or a Samaritan), the anything and everything is acceptable, eat, drink and be merry style of eating. There is a distinct separation between the person who is eternal minded and the person who is "live in the now" minded and driven by appetite and cravings. To eat at a table with people who are eating foods that are not acceptable to the Hebrew DNA results in the essence of those unacceptable foods transferring energetically to the one who is refraining (the Hebrew). The body receives the signals from the foreign food source. Those with the fully functioning Hebrew DNA (aka Spiritual Royal) will pay a price in sleep quality or digestion when their table is shared with outsiders who consume the unacceptable foods. Unacceptable for acquiring or maintaining Hebrew DNA that is. The body is very intelligent.

Continuing from verse 32 above, *Genesis 43:33: They were seated before him in order by age, from the firstborn to the youngest. The men looked at each other in astonishment.* (CSB)

It appears that seating arrangements at a table have a significance, otherwise there would be no mention of the order in which they were seated. I cannot speak for families outside of the United States but most families living in the United States may have a specific seat they prefer at their family dining table, yet not necessarily being seated in an age-related order.

Tables are not only a sign of social status but can also reflect a place where formal agreements are entered, and legal documents are executed. With this in mind, what message are we sending to the cosmos when we sit in a restaurant full of tables amongst various classifications of persons, particularly those we do not know? Are we making an agreement with the other patrons whom we do not know? And what message is being sent to those who lack the means to provide food for their table? Maintaining a level of privacy in your personal life has great value and what is done in public is not always beneficial.

1 Corinthians 11:22 and 34: Don't you have homes in which to eat and drink? Or do you despise the church of God and humiliate those who have nothing? What should I say to you? Should I praise you? I do not praise you in this matter! Verse 34: If anyone is hungry, he should eat at home, so that when you gather together you will not come under judgment. I will give instructions about the other matters whenever I come. (CSB)

The first item that needs to be cleared up is what the word Church means. Church: medieval Greek *kurikon*; Greek *Kuriaon* (*doma*) 'Lord's (house);' from *kurios* 'master or Lord'; compare with '*kirk*' – circle or company or assembly. The word lord is comparable to say, a driving force, something that motivates you.

Church is a name for a collective group of people that live a same or similar lifestyle and share the same beliefs or goals. A good example for this would be the Amish community. Their community could be called a "Church" of people. Another example would be wheat farmers. Collectively as a group they could be called a Church of people. They have the same vision, the same goals and their daily, monthly, yearly cycle of work-related events are generally the same. The ancient meaning of church has nothing to do with weekly meetings, sermons or denominations.

These verses are clear, when we eat, we should be at home. Consequences (judgment) will ensue when meals are eaten away from your home dining table and in public amongst outsiders. This insight may also make a person think twice before participating in any public or church dinner! That would make headlines: "Attend a food focused function and leave with a judgment attached to you." That judgment would be in the form of a consequence that can present itself in numerous forms, particularly those that involve health.

Meals are to be eaten indoors, not in the environment where animals leave their messes. There will be times during travel that we must eat away from home. How these situations should be properly addressed will come with time and experience. Before drinking from a glass in a restaurant, take your finger and write your name on the glass. To use a drinking glass, even though washed, that has been used by multiple people will eventually result in a virus. Personally, I have traveled with my own food and eat at my accommodations. If travel prevents me from taking food with me, I locate a grocery store to purchase what I need and eat at my accommodations outside of the presence of strangers. There will be times when formal dinners served by properly dressed and gloved individuals can be enjoyed.

Exodus chapter 12 addresses the instructions for the Passover. The Passover is indicative of specific eating regulations that will ensure the blood remains free of affliction and the death angel of disease and decay does not pay you a visit. The highlights of the Passover meal is a guide for how meals, particularly those that include meat, must be eaten. The yearly marker of Passover is a reminder of the rules that must be applied monthly throughout any given year.

<u>Exodus 12:11</u>: Here is how you must eat it: You must be dressed for travel, your sandals on your feet and your staff in your hand. You are to eat it in a hurry, it is the Lord's Passover. (CSB)

Dressed for travel is a reference to how one would be (or should be) dressed when in public with shoes on your feet and proper attire in place. Meals should <u>not</u> be eaten while in your night clothes, bathing suit, tank top or shorts, and so forth. Proper hygiene is required, and proper dress is also a must. Dress as though you are participating in a nice dinner, not necessarily formal but business-like. This practice could be classified as Royal etiquette.

Formal Dining

Eating out for a formal dinner or special occasion can become costly to the electrical storehouse in the body. Meals must be served by someone wearing a glove on their right hand, no glove on the left. Proper glove etiquette neutralizes any electrical energy that would come from the kitchen or the server. Double layer jackets (jacket w/vest) should be worn by all servers. Left hand serves butter; right hand serves the plates. Toddlers are to be seated and eat in a connecting room, not in a formal dining area.

When attending dinner parties, the less people seated at one table the better. Four to six persons per table is recommended. Tables that seat larger numbers produce an abundance of conflicting energies as people eat. It is best to limit conversation during any meal. You may literally end up eating your words and the words of others as you chew and swallow your food. Specific seating arrangements are required for banquets with officials or attendees that hold some form of authority being seated at tables in the center of the room.

Meals

Proverbs 23:20-21: Don't associate with those who drink too much wine or with those who gorge themselves on meat. For the drunkard and the glutton will become poor, and grogginess will clothe them in rags. (CSB)

Gluttony: Habitual greed or excess in eating.

Before being seated for a meal, hands should be washed under running water and the mouth is to be rinsed with warm water. If you have been outdoors, washing the face with a clean, cotton wash cloth is beneficial, no splashing water onto the face. Mealtimes are connected to the position of the sun, activities in the atmosphere, and planet alignments. The Heavenly Gases can become disturbed or out of balance when meals are not eaten at the appropriate time. For the maximum benefit, eat meals at the designated times when possible. Eating outside of mealtime guidelines can lead to an oily face, fatigue, interruption of the Heavenly Gases and digestive disturbances.

The body will play games with appetite and cravings. Cells recall what has been eaten and when those cells are awake, they can send a message to the brain that a particular food is what you need/desire/crave. Certainly, there are times we need a source of nourishment but maybe this required nourishment needs to be viewed with a different set of eyeglasses. Nourishment does not necessarily mean fill the stomach until it ceases its grumbling. Nourishment means giving the body what it needs to function properly. The body may be lacking in something outside of a vitamin or mineral when it sends a signal of hunger. Nourishment: food or other substances necessary for growth, health and good condition.

Large events with catered meals, church dinners, and so forth come riddled with various energetic debris! To eat food prepared in a kitchen that has little or no supervision with respect to cleanliness and unfamiliar people cooking or baking the food that you are consuming can have some energetically damaging effects. This rings true with restaurant foods as well, even though monitored by state regulations. Some insight gained from these various events or establishments that provide meals has become quite unsettling. From the foreign debris that can easily accompany processed foods to the people who prepare those foods not using proper hygiene, foods consumed outside of your own kitchen can have impact on the body in general, not just the stomach. Here's some things to consider: Fluid in the brain, spine and lymph can become contaminated or stagnant, resulting in excess mucous production. Mucous is a normal response to infection. Are there medically unidentified infections taking up residence in the brain and spine due to contamination encountered in public eating situations? The increase in people diagnosed with Alzheimer's or Dementia raises the question as to whether restaurant foods or a meal delivery service has contributed to brain issues.

Breakfast

The most beneficial time to eat breakfast is an hour after sunrise although there is no considerable harm in eating any time prior to 10:00 a.m. Bread is to be eaten at breakfast, providing the chemical activity in the body for production of new cells. Many breakfast cereals will advise that their product contains gluten but the measure of active cell-producing influence the gluten has in that manufactured dried cereal would be minimal, if any. Gluten protein must be consumed the day it is activated to receive the benefit needed. Freezing a gluten activated product will stop the decline of vitality in the gluten.

A good alternative to dry cereal is a fresh made pancake. Pure maple syrup, no butter and a few sunflower seeds or pumpkin seeds makes for a beneficial breakfast. Dried cereals and mass-produced cereal products have little or no value. Besides, dry breakfast cereals can contain bugs! Ick. The important components to be aware of are the Heavenly Gases. The nutritional panel on the side of the box means nothing at this point.

Fresh fruit can be had periodically throughout the day, avoiding the hours of noon to 3:00 p.m. I try to keep any snack at a 200 calorie limit. Any fruit should be eaten at least an hour and a half away from any bread or grains. The heat of the day removes the vitality that is necessary for the proper digestion of food and interruption of the internal Heavenly Gases can occur.

Lunch

Most lunch breaks are between the hours of 11:00 a.m. to 1:00 p.m. In the book Living by the Light of the Moon, I expand on this topic and list Scripture that sheds light on the activity of the atmosphere that influences the body, specifically during the hours of noon to 1:00 p.m., and even until 3:00 p.m. After having breakfast between 7:30 a.m. and 8:30 a.m., I found that eating a snack, usually of fruit with a limit of 200 calories, between 10:30 a.m. and 11:00 a.m. provides satisfaction to the stomach with little or no problem in waiting until 3:00 p.m. to eat something more filling. Considering the intensity of the sun and possibly other planet activity (interruption of Heavenly Gases in the atmosphere) during these hours of the day may provide us some answers.

Abusing the lunchtime restrictions can result in genetic disruptions that lead to earaches in your descendants. This type of activity would involve either excess fluid accumulations in the head and sinuses or a component of pressure, or both. Sandwiches are related to the development of asthma, memory issues, Reynaud's, pain in the neck/shoulder area, breast and lung tissue irritation, breast, pancreatic or lung cancers, fatigue, and memory disturbances. Sandwiches cause electrical interruptions in the body and water retention. Most sandwiches are made with a yeast bread and yeast can be a troublemaker in the health department. There is a connection between eating bread prior to sunset and grief.

II Samuel 3:35: Then they came to urge David to eat bread while it was still day, but David took an oath: "May God punish me and do so severely if I taste bread or anything else before sunset!" (HCS)

This verse has two interpretations 1) when grieving, do not eat until after sunset, with an emphasis on bread; 2) to eat bread (outside of for breakfast) prior to sunset can create circumstances of punishment and result in the emotion of grief. The punishment could be in the form of digestive upset, blood sugar irregularity, or lack of the Heavenly Gases during nighttime when the body is doing its Heavenly Housekeeping, which would result in an entire host of health issues.

Dinner

An evening meal should be eaten dependent upon the hour of sunset during each season. Meals eaten late afternoon prior to dinner do not have the same benefit as they would when eaten at the appropriate times. The most important reference to an evening meal is in what is been given the name of The Last Supper. Matthew 26:20 states: "When evening came," not giving a specific hour but leaving an impression that depending upon the season the time in which evening comes will vary. After sunset yet before night sky is considered evening and is the best time to consume your dinner/supper meal. Bread was the main course for the Last Supper. Eating vegetables for dinner can have the same result as eating beans. When eating vegetables, they should be eaten earlier in the day as they have a reputation of interrupting sleep when eaten in the evening.

Holiday Meals

This subtitle supports *Proverbs 23:1-2:When you sit to eat with a ruler, consider carefully what is before you; and put a knife to your throat if you are a man given to appetite.* (NKJV)

Buffet style eating and holiday meals qualify as gluttonous activity and according to Proverbs will land a person in weakness (poverty) or even death. Gluttony also will result in a large waistline. Whether gluttony is a personal issue or an issue of an ancestor, gluttony is a player when the waistline becomes out of control and the body becomes weak. Physical weakness is a sign of damage to chromosomes.

Numbers 11:33-34: But while they were gorging themselves on the meat – while it was still in their mouths – the anger of the Lord blazed against the people, and he struck them with a severe plague. So that place was called Kibroth-Hattaavah ("graves of gluttony") because there they buried the people who had craved meat from Egypt. (NLT)

We need to closely consider the date and times of certain holiday meals within the United States. I cannot speak for countries outside of the U.S. The traditional U.S. Thanksgiving Day can fall within the Waning Moon when meat and dairy products are best to be left out of the menu. Thanksgiving meals are generally centered around large portions of various types of food, which would cause the holiday meal to qualify as a gluttonous occasion. Meals during the Christmas holiday are running a close second in the gluttonous event category. When Scripture instructs us to avoid being a glutton, why are these food focused events so popular? Part of the reason is because sermons inside churches have taken the message off course.

A brief search into the history on the development of the U.S. Thanksgiving Day, President Abraham Lincoln was impressed by the writings of Sarah Josepha Hale, who wrote children's poetry (i.e. Mary Had a Little Lamb).

The final Thursday of November was proclaimed as the national day for Thanksgiving in 1863. It appears there was no educated consideration given to the selection of the day for Thanksgiving prior to signing this proclamation. Here again is a situation of a person in a position of authority initiating activity that can result in a degree of ruin. The concept of Thanksgiving Day is not harmful but the date or even season in which it is recognized can have a negative impact on the body. Might I suggest that someone with the necessary knowledge, meaning knowledge of the orchestration of the cosmos and the influence it has on eating, reset the date for Thanksgiving Day for the United States? I'll also suggest that turkey not be the focus of the holiday meal. Eating turkey can result in issues all its own, such as genetics for being lazy (tryptophan).

Thanksgiving Day November 2023 brought about events in the Heavens that caused star gazers to be summoned and give attention to Saturn and Jupiter, two planets with a high concentration of helium and hydrogen. Obviously, some Heavenly Energy going on there that likely holds a message of some form. Couple the Saturn and Jupiter debut with Thanksgiving Day celebrations and there could be a hidden message from the Heavens in this cosmic event. Here's that mysterious God language again! Could Thanksgiving Day feasting, or even feasting at any time, result in excess helium and/or hydrogen in the body? It appears someone in the Heavens was waving a flag to get Thanksgiving participants to pay attention. It is beginning to look as though people are gassing themselves to death. Do the words Nazi Germany and Hitler ring any bells? The events of the 1940s may have been a warning sign for what is taking place in people today in response to the gluttony.

Large family gatherings are not only a lot of work and quite time consuming if you are the cook, but also have downfalls energetically. Eating is a sacred event and gathering with numerous people at a time for eating a meal will not only influence those eating the meal but the infants and children that are present as well. Babies will cry when the atmosphere is contrary to what it should be during a meal. Babies have a very sensitive signaling capability and when there is an overload of vibrations permeating the airwaves that are contrary to the otherwise harmonious atmosphere the baby is accustomed to, the baby will cry. Expose the baby to these types of environments enough times and the baby could develop a habit of crying during mealtimes. Celebratory meals can result in ringing in the ears (tinnitus), a result of the noise from conversations that accompany large gatherings. Limited conversation should be had at a dining table.

The word feast comes from the Latin root festa, meaning joyous. This brings up the question as to whether Scriptures that reference feast are speaking of a time of joy, not a time of eating. In some instances, the mention of a feast could be a marker for being joyous or thankful, to show appreciation for something in particular. Scripture mentions feasts being on specific dates according to the moon phase. Feasts were generally held on or near Full Moon or New Moon and not all feasts were acceptable.

Isaiah 1:14:Your New Moons and your appointed feasts my soul hateth; they are a trouble unto me. I am weary to bear them. (KJV)

This verse is saying how the Star Dust (the electrical charge in the plasma which makes up the Soul) is influenced by certain feasts, giving rise to the Soul becoming compromised.

Psalm 81:3-4: Blow up the trumpet in the new moon, in the time appointed, on our solemn feast day. For this was a statute for Israel, and a law of the God of Jacob. (KJV)

A trumpet is symbolic for an announcement, to be aware that something is forthcoming. New Moon marks the beginning of a new cycle. Are we being told to be joyful at the coming of the New Moon instead of sitting down at a feasting table?

Toxic Tables

The atmosphere generated from idle talk and gossip, particularly while seated around a dining table, creates a highly toxic environment. This environment breeds, spreads, and grows to the point it damages the heart muscle, valves or even other parts of the body. The words spoken while seated at a dining table are what you will eat.

Conversations that spinoff of personal opinions, selfish-ambitions or emotions can result in death, not only spiritually but also physically. Many physical heart issues are a result of exposure to this toxic environment, whether you are part of the conversation or not. A root of bitterness, envy, jealousy and strife take hold and can easily become attached to a food you are eating at the time.

Proverbs 18:21: Life and death are in the power of the tongue, and those who love it will eat its fruit. (HCS) (emphasis added)

Acts 8:23: For I see you were poisoned by bitterness and bound by iniquity. (HCS) (emphasis added)

Hebrews 12:15: Make sure that no one falls short of the grace of God and that no root of bitterness springs up, causing trouble and by it, defiling many. (HCS) (emphasis added)

James 3:6, 14, 16: And the tongue is a fire. The tongue, a world of unrighteousness, is placed among the parts of our bodies. It pollutes the whole body, sets the course of life on fire, and is set on fire by hell. v. 14: But if you have bitter envy and selfish ambition in your heart, don't brag and deny the truth; such wisdom does not come from above but is earthly, unspiritual, demonic. v. 16: For where envy and selfish ambition exist, there is disorder and every kind of evil. (HCS) (emphasis added)

CHAPTER X
WHO SHOULD WE BELIEVE?

Humanity has advanced to a position of trusting in those who have a line of alphabet behind their name. Does humanity need to stop and reconsider how much truth and measure of totality are in all of the reports, scientific studies, popular voices or promotions provided to the public? Consider how nature works, then think about the influence of nature's opposites: chemicals, manufacturing, preservatives, flash freezing, drying, and the list goes on. The view becomes a little clearer with respect to the fact that maybe those advertisements and promotions printed or seen on TV are not telling the "whole truth and nothing but the truth." There just might be some loopholes, sinkholes or deep caverns that have been glazed over, hiding a bit of the truth that all should be allowed to see.

The energetic influence of a particular item, whether food or the packaging it comes in, will relay signals to the body. While this energetic language can be new and quite foreign to some, it should be considered when deciding whether a product is safe to consume. For example, eating leftovers can result in the body hanging onto things it needs to eliminate such as toxins. Could this also be a cause of constipation? What about eating dried foods? Does this relay a signal to the body that would result in dry scalp or skin, maybe even wrinkles? Many foods when dried shrivel and become wrinkled in appearance. The list of signal – response activity can go on and on. These signals and their result were not considered by those who approve and label items for consumption. This is a good place to throw in food manufacturers emphasizing holiday meals and celebratory events that include their specialty dish. How much consideration is given to the impact their popular food dish would produce? None, other than satisfying the cravings that are classified as gluttony.

To ignore or disobey the guidelines for consumption of food will lead to contamination in the plasma resulting in the lymph system becoming overloaded and sluggish. This causes the lymph system to be unable to sufficiently remove toxic debris. Tonsils are a part of the lymph system and tonsilitis, strep initiated cases or not, is far too common. This pattern of incorrect consumption of food can lead to infections in the blood that will initiate a whole host of health issues. It is clear, foods hold the key of judgment that comes to the body. The water (plasma) in the river (veins) must be clean or the people die.

<u>Mark 6:21-23, 27-28</u>: An opportune time came on his birthday, when Herod gave a banquet for his nobles, military commands and the leading men of Galilee. When Herodias's own daughter came in and danced, she pleased Herod and his guests. The king said to the girl, "Ask me whatever you want and I'll give it to you." He promised her with an oath: "Whatever you ask me I will give you, up to half of my kingdom." <u>Verse 27-28</u>: The king immediately sent for an executioner and commanded him to bring John's head. So he went and beheaded him in prison, brought his head on a platter, and give it to the girl. Then the girl gave it to her mother. (HCS)

There's a lot going on in these verses. First is the reference to banquet and birthday that would involve much food. Galilee means revolving or reoccurring (anniversary or yearly event such as a birthday). An oath is a formal promise, yet we are instructed in James 5:12 to let a "yes" be "yes" and a "no" be "no." Herod has quite a lengthy definition, which includes the words homeless, wanderer, fugitive and wild ass. To be "homeless" in the context of these verses means to be away from the Heavenly Home lifestyle, the ways to live that are instructed by God, or the cosmos. This oath given by Herod, who is in the position of authority, attached 1) reoccurring food focused events result in affliction to the head/brain; 2) water (plasma) issues, evident in the character role of John the Baptist (down under, or inside the body + water). It is a violation of the natural order of things (life) to involve yourself in gatherings focused on yearly anniversary markers coupled with a feasting table that accompanies numerous food dishes. This activity initiates problems with respect to the plasma, particularly in the region of the brain. Yes, the gut and the brain speak to each other. The plasma contamination must be addressed prior to being granted any form of rescue from a physical disease or affliction; John the Baptist comes first, then Jesus. This water/plasma component causes the Heavenly Gases needed for the health and vitality of the physical body to become out of balance resulting in diminished function, giving rise to health disruptions.

Food Preparations

With women moving into the work force, meal preparation took a turn toward anything quick and easy. There is less time for grocery shopping, meal planning and preparing a meal when the chef of the house is away from home for 8-10 hours of the day. Add a few children to the mix and there is even less time. These situations produced the slipper slide to the bottom of the necessary food intake requirements with many meals being made from shelf-stable products versus fresh foods that hold the value necessary for the chemical gas production in the body. All that is accomplished is filling the stomach with worthless "stuff" that takes energy to push through the digestive system. By this point you, like me, are thinking it is not so easy being a human! The following helpful hints will assist in relieving the burden on the body when it comes to eating.

Standing directly over a stove where heat is rising toward the head and face can be damaging to the cells and will deplete the Heavenly Gases inside the body. Limit standing near the heat of a stove, oven or any type of open flame and certainly do not allow children to stand near the heat. Make sure the stove surface and burners are clean prior to beginning your meal preparation. Avoid no-stick coating and cast-iron cooking pots and pans. No-stick coatings eventually flake off and can end up in your food. Cooking with cast iron can elevate the iron level in the blood. Copper chore girls work best for those tough cooked on or baked on items and a plant-based dish soap is a better choice for protecting your skin and cells from harsh chemicals. No dish detergents with degreasers. They harm the cells.

Avoid using a microwave. Eating food that contains those tiny electromagnetic radiation particles used to cook your food will rob you of your energy and can result in alterations in vision. The message the body receives from food cooked on a gas or charcoal grill can result in the interior temperature of the body becoming elevated and if that message records in the DNA, the heat issue could be continual.

Large spoons and ladles are for stirring food while it is being prepared or during the cooking process. Tasting or eating foods with large spoons will result in an increase to the size of the mouth. Forks cause a response that increases fluid production in the body.

I realize how convenient plastic bags, plastic wrap, and aluminum foil can be but there are health hazards when these items contact food. Once plastic is in the body it will migrate to the brain and eliminating plastic from the body can take as long as 17 years. What is this plastic doing to the brain? Could this be the childhood ADD or ADHD that came about in the 1980s? Plastic drinking straws made their debut in the 1960s, replacing the paper straws, creating just enough time between introduction of plastic straws and the discovery of ADD. Something to consider.

Food preparations can fall into a category of being a craft. Mixing numerous items together to create a particular dish not only involves numerous ingredients but also involves mixing and stirring. Placing ingredients in a pot, stirring the contents while it is positioned over heat is what is commonly seen in depictions of a witch brewing up a potion. There must be something more to this collection of activity than what is commonly known. Old folklore or an old wives' tale would spill the beans on this subject. Aside from the image of potions, this theory tells us that there is an influence to not only cooking up a multi-ingredient dish, but also an influence received from the product cooked and the cook who is stirring the pot. It is unknown what emotions or life situations the cook who made your dish has experienced, or what words were spoken while they made your meal. Energies from the cook or other persons who handle your food prior to you eating it, will transfer to the food.

Food Selections

Fresh homemade yeast bread, unleavened flatbread or flour tortillas are staple foods. Bread is made with flour, water, yeast, a pinch of salt and olive oil. Tortillas and flatbreads are made from flour, olive oil, pinch of salt and hot water. Organic wheat flour has the required nutrition, and the oil and water are avenues to activate the gluten proteins in the flour. I recommend using Celtic salt or Redmond's salt, both have beneficial minerals. Standard table salt is overly processed and contributes to the hair turning white. Table salt is good for mixing with baking soda and using as a scouring powder for cleaning sinks and tubs.

Fruits, vegetables and herbs should be fresh; home canned or frozen are second options. Dried herbs have no value other than adding flavor to your dish. Do your best to eat fruits and vegetables in their season. Having access to manufactured and processed canned foods has contributed to the diet downfall allowing for consumption of fruits and vegetables during all seasons.

Just to add a little zest to the next bite of processed food you put in your mouth, has the thought ever crossed your mind of how likely it is that processed foods contain fragments of insects or rodent droppings? Massive industrial buildings that mechanically process foods day after day can only be cleaned to a certain degree. Mice and other crawling critters find their way into the machines and leave their tracks or remnants in places beyond where the eye can see, or the cleaning crew can reach. This is not an appetizing thought but a very likely scenario. Yuck!

Restaurants and Cafeterias

Having covered this topic to some degree in Chapter IX, Setting the Table, I will use this subtitle to expand on a few points.

High-end restaurants are a safer choice when there is a need to eat away from home. While some of the common go-to eating establishments may have the best dish for satisfying your tastebuds or cravings, they may not always be the cleanest or prepare their menu selections with the highest quality ingredients.

Restaurant (or Cafeteria) food is not compatible with Hebrew DNA if you desire to avoid damaged DNA. Healthy DNA reads restaurant food as being foreign, a food prepared by someone you do not know and more times than not, the food is not freshly made. When the DNA is damaged it causes an interruption in the copying and signaling processes of the cells. This could be compared to a bent spoke on a bicycle wheel. The bicycle still moves forward but the ride may not be smooth and will cause wear on the tire. Eventually something will fail.

1 Chronicles 12:38-40: All these warriors, lined up in battle formation, came to Hebron wholeheartedly determined to make David king over all Israel. All the rest of Israel was also of one mind to make David king. They spent three days there eating and drinking with David, for their relatives had provided for them. In addition, their neighbors from as far away as Issachar, Zebulun, and Naphtali came and brought food on donkeys, camels, mules and oxen – abundant provisions of flour, fig cakes, raisins, wine and oil, herds and flocks. Indeed, there was joy in Israel. (CSB)

These verses reference food being made and provided by relatives. Issachar, Zebulun and Naphtali, although referring to locations in this verse, were distant relatives of David.

The energy from restaurant food interferes with women's menstrual cycles that can lead to infertility and contribute to painful teething episodes for infants and children due to cell signal interruptions in the gums/mouth. Could a similar tooth pain issue in adults be a result of eating restaurant foods? Teeth are very energetically sensitive.

Most restaurants conform their menu from mass produced canned foods. The past few years have brought about a few restaurants that provide fresh and organic items on their menu. Ordering takeout from a restaurant has a level of risk as well. While you may not be sharing a dining area with strangers, you are still consuming food prepared by strangers.

Open buffets or potluck meals have a whole host of energetic disturbances all their own. You don't know what has gone on in that area before you serve your plate, nor how many hands have touched the serving utensils. Your inner power is easily diminished in these situations.

Thinking of the environment of a restaurant, genetics for leg pain or leg injuries can take seed if you have an ancestor who spent long hours on their feet waiting tables. I doubt there has been any form of survey done on the following apron theory, but I do know that white is not a compatible color for anyone at any time if you want to maintain chakra (wheel) energy. White washes out color vibration. The chakras in the body need color to maintain function and an energetic setback can take up to three months to recover. A hemline will also influence various parts of the body. Having a hemline of a white apron landing at mid-shin, with a tie around the waist causes energy to collect in the region of the hips and eventually the derriere becomes large. An apron with a loop that goes around the neck will result in neck and/or shoulder pain. Colored aprons may be a better option. Black should be reconsidered as well unless you are a male in a black suit with a pastel shirt and colorful necktie. Black shirts should be avoided. Wearing black at any time can be tiring.

While it is fun to sit outdoors at your favorite restaurant while eating your meal, it is not energetically advisable. Various animals, including rodents, frequent restaurant outdoor areas looking for a few crumbs and the deposits they leave behind, even though visibly cleared away, leave an unseen message for those seated where they have frequented. Indoors is the proper place to eat a meal.

As you read through the next few chapters on specific foods, note that much of the influence that comes to the body through food takes place in the plasma. Plasma is the river that runs through the body. The goal for living healthy is for the plasma to be the River of Light and in order for the plasma to be the River of Light it must be clean.

I will wade through food groups and share what I have learned about their benefit or detriment to the physical body. I have had several years of practice with dodging specific foods for various health reasons and have numerous extended family members and friends that are challenged in the same arena. Stepping back from an individual, food specific view and taking on a grand overall view of the common food challenges, a few keys to the internal dilemma being experienced were discovered. The mind must step back to times long before Food Pyramids and diet and nutrition guidelines, which were developed in response to the lack of proper food being consumed in the first place. Food Pyramids and nutritional guidelines have obviously failed us. The body is an amazing machine and when given the proper fuel it will purr like a kitten!

CHAPTER XI
BEVERAGES

Is it possible that the human thirst response originated from the consumption of animal flesh or animal products? Cows (beef), being a more popular consumed flesh, drink a significant amount of water each day. Scripture describes the origins of Hebrew DNA as having an ability to hydrate itself by consuming little or no water, or even having a desire to drink water. The Living Water Scripture speaks of would be a result of the Heavenly Gases, and any actual liquid consumed would be from fresh fruits or fruit juice. Animals naturally drink water and would have a thirst response built into their DNA.

Hot beverages need to be warm, not excessively hot. You should have the ability to hold the drink in the mouth before swallowing. Drinking hot beverages sends a message to the body to maintain a temperature that is too warm and can result the plasma producing harmful condensation. The opposite influence can be said for iced or frozen beverages, resulting in the body receiving a signal that it needs to remain cooler than it should for good health.

What is taking place inside the body when excess fluids are being consumed? If the Heavens provide the gases necessary to produce the hydration (Living Water) we need, what are the glasses of water, cola, coffee and so on doing to the Heaven Hydration? Would the excess liquid cancel out the benefit that would otherwise be present in the body? What if all the excess liquid takes up residence in the tissues or lungs and produces condensation that eventually springs up as bacteria or inflammation?

John 4:14: But whoever drinks from the water that I will give him will never get thirsty again. In fact, the water I will give him will become a well of water springing up in him for eternal life. (CSB)

It appears people have marched themselves into a muddy marsh of health problems by over-consumption of liquids. The liquid itself is not the only detriment at play, many items classified as refreshing beverages come with a price to the internal signals that orchestrate the function of the body. Artificial flavors, color and even carbonation all have a negative influence on the blood. Sugary beverages take up residence in the genetics, resulting in genetic imprints for glucose issues and obesity. Yes, people who are not properly connected to the Heavenly Hydration may need some form of hydration until their body begins to operate through the processes that naturally produce the hydration needed. I chose to eat fruit and drink fruit juice for my hydration. My body has graduated into production of the hydration it needs so I have been able to reduce the fruits and fruit juice I consume each day. I rarely drink water and if I do, I drink maybe 2-4 oz. I can go weeks, even months without water.

Alcoholic Beverages

Scripture references wine on numerous occasions but the root meaning of the word wine had origins not commonly known. The etymology of the word wine has numerous references to various countries. Two definitions are: vino or vine (a reference to the grape vine not fermenting the grapes); and to bend or twist. The word wine in Scripture is a reference to Soma, pressed juice, the blood of the grape. Notice the descriptive words "immortal" and "light" in the verse below. One verse within the Vedic Hymn entitled Soma states:

"We have drunk Soma; we have become immortal; we have gone to the light; we have found the gods. What can hostility now do to us, and what the malice of mortal man, O immortal one? (The Golden Book of the Holy Vedas)

Eating or drinking something fermented creates gases inside the body that conflict with the Heavenly Gases and in turn causes contamination to the Hebrew DNA. Wine or other fermented drink would not have been a common drink for those who desired to advance into a position of Spiritual Royal (those who have "Christ" – blood without contamination). Wine, like many other foods and drinks, travels through the generational lines and can stir up many health issues. Wine results in genetic imprint for a pudgy or puffy appearing face.

I found the following fact interesting, the word wine has a Latin root of vinum, meaning to minister to one's bodily wants. Why not just say lust of the flesh? And this is what the church is telling us to drink as part of communion? Satisfy the "wants" of the flesh and you could be eating/drinking poison. Need I say anymore?

Alcoholic beverages cause excess fluid in the body and hinders communication with the cosmic signals. Dampness in the body is a result of drinking beer and liquor. Descendants of brandy drinkers prefer an early to bed routine, while descendants of whiskey drinkers entertain the late-night hours.

Coffee

Oh the coffee bean! They smell so good when they are being roasted but what is happening inside the body once brewed and consumed?

Looking beyond the sleep interruptions coffee can cause, coffee will weaken your inner power by acting as a sponge that sucks up all of the power inside you, specifically in the solar plexus region, the region of the belly. The influence of coffee will also travel through the genetic lines and all of those cups of coffee grandma and grandpa drank could cause the descendants not only sleepless nights but other health disruptions. Balding and thinning hair are a sign of too much coffee consumption in the genetics. These are some of the more easily addressed issues but what is this bean doing to our DNA strands?

Think about the oily residue left by coffee on a coffee pot. That oily residue can, and eventually will coat the lining of the arteries. Is this a contributor to clogged arteries commonly experienced? What about the lining of the intestines? Coffee will stick to the lining of the intestines and could result in a reduction in the process of eliminating waste. The ganglia cells, the cells that push waste through the intestines, become hindered.

No doubt, coffee interferes with the electrical function of the body and increases internal heat, not to mention coffee is often full of mold or chemicals.

Coffee is a contributor to the hyped-up lifestyle we see today. Nervousness, jittery responses the body displays pour out into daily life and could be a root to the hyperactivity in children. This hyped-up lifestyle interferes with the body's natural process of Heavenly Gases. Revving the engine at too high of an rpm results in the unnecessary burning of gas.

Drinking any form of hot beverage from an insulated cup produces a bubbly sensation in the stomach developing into belching or flatulence, both signs of an interruption in gases. Styrofoam is made of toxic chemicals and should not be used for consumable food or drink.

Fountain Sodas

Carbonated drinks were covered a few paragraphs ago so let's now consider the artificial color, flavor and aluminum containers. It is well known that some artificial colors can cause health issues, especially in children. Is that not a clue that they are a danger to the blood? Some of the ingredients and chemicals may not cause an immediate reaction, resulting in dismissal of any symptom being related to the contamination encountered through the drink consumed. It may be years of accumulation or genetic contributors that pile up before there is a disruption in health and the answer to how that specific health issue came about would be "we don't know." There are ten thousand combinations how aluminum cans, colors and flavors can develop into a symptom and tracking them to any one illness, ache or pain would be nearly impossible. This does not mean there is not an avoidable cause. This is a good example of where the language of the Universe would need to be applied. If it is not present in nature, don't eat or drink it. There are Universal laws that govern these situations (and many others) we just need to learn to apply the language. Having had the personal experience of heavy metal toxicity, there is no form of aluminum that should contact food or drink.

Coke, which contains phosphoric acid, can create a genetic imprint for obesity and cause an instability in the brain activity that controls the body's natural ability to cool itself. Coke may not be alone in these situations, although we can eliminate caffeine as being the sole culprit since these alterations in the body haven't appeared in other caffeinated beverages, to date. Dr. Pepper, a childhood favorite, contains another combination of ingredients to be questioned. Dr. Pepper contains phenylalanine. Do we even know what we are consuming when we drink these products? I have no doubt that artificial sweeteners combined with carbonation can escalate a person's chances for blood infection or cancer.

What about sports drinks? Sugars, food coloring, artificial flavors yet they are touted for an ability to replenish hydration in the body. They may temporarily do so but what else is going on when these concoctions are consumed? And for how many generations does their influence remain in the body? It is becoming evident a whole lot more studies need to be conducted that are not under the influence of the company who manufactures the product.

Juice

Concord grape juice is the Soma the Vedic Hymns reference. It is also the wine referenced in the Holy Bible. Grape juice has a cleansing and energizing influence. Concord grape juice may be consumed throughout the year, assisting in the cleaning of the plasma and removal of excess water. Concord grape juice could be put in a category of blood tonic, providing a source of life energy.

Contrary to popular belief, fruit smoothies should not contain greens or added protein powders. Mixing numerous fruits and vegetables into a drink has a negative impact on the blood. This type of drink could qualify as gluttony, giving the body too much information to process at one time.

Beet juice will cleanse the bowels so go lightly with this juice.

Citrus juices, specifically orange juice does not seem to mix well with Hebrew DNA. Overall citrus can create a histamine response in those with any form of allergy.

Tea

Tea should be used for medicinal purposes. Too much tea can generate too much internal heat. It takes a lot of energy for the body to process tea. Rooibos tea has high antioxidants and can be used for medicinal purposes.

Water

Cells in the body respond to water. The sound of water or touching water will have an impact on the cells and the cells in turn take an action. Similar to hearing water running may cause a person to have an urge to urinate. This water activation scenario rings true in various categories of life in general, from washing the hands to the sound of the washing machine or dishwasher running. Too much water activated response in the body can lead to the plasma becoming challenged in doing the necessary cleansing. Water needs more respect than what it is generally given.

While the Israelites were in Egypt they had food and drink to their fill. Once they left Egypt in response to God's command, they discovered that the land (situation) they were to live in did not consist of the abundant food and drink they had grown accustomed to.

Numbers 20:5: Why have you led us up from Egypt to bring us to this evil place? It's not a place of grain, figs, vines and pomegranates, and there is no water to drink! (HCS) (emphasis added)

The craze to drink half your body weight in ounces of water each day has done its share of contributing to a water-logged body. The body was not designed to metabolize water and over consumption, such as suggested in various reports or studies, can lead to excess fluid accumulation in the spine, brain and other areas of the body. Swollen limbs may not just be the result of consuming salt, it may have a connection to over consumption of water. It appears that water consumption is intended for our 4-legged friends.

Excess water has an influence on the naval area, the meridian that runs up the center of the body up through to the crown of the head. Excess fluid in the body can mean pollution in the plasma. Signals or vibrations encountered during the normal day activities can record in the plasma. When fluids are in abundance, the body cannot eliminate the signals, and the signals begin to interfere with the normal function of cell communication. This results in the internal electrical power becoming hindered. Chronic or spontaneous cough is a result of excess water.

Water packaged in plastic bottles has created an avenue for consumption of plastics. No matter the grade of plastic, once water is placed in a plastic container the water will absorb and record the negative output of the plastic. Any man-made product that contains chemicals will give off a negative vibe. Water is not the only consumable that is delivered in plastic. All plastic containers that are used for food or drink will deliver a level of plastic into the body via the food or drink they contain through a means of vibration. Consumption of plastic will cause havoc on the brain and can turn the vertebrae in the spine to a rubber or silicone-like substance and could result in years of attempts to detox the plastic out. Any condensation in a water bottle is a sign of bacteria, toss it out!

Consuming unfiltered city water has risks, one being the influence fluoride can have on the function of the thyroid. Reverse osmosis water removes a large percent of any contaminates in water. For cooking or consumption, reverse osmosis water is recommended. It is best to avoid drinking water with a meal, particularly with breads.

Ever drink coconut water? I used to like to eat coconut now and then, but I have learned to remove it from my diet. I was never fond of coconut water but did try it only to find that it irritated the lining of my stomach.

Carbonated water influences the body like being in a state of intoxication. The carbon dioxide gas used to make the bubbling activity alters the blood in a way that slows down or hinders the body's internal functions, with emphasis on the pancreas. It appears these seemingly harmless bubbles are throwing off the natural balance of the Heavenly Gases that are necessary for the blood/plasma to do its job. More is going on here than what is being reported by the experts.

A sip or two of warm water with lemon and honey is beneficial when experiencing a stuffy head. Go easy on the lemon.

Beneficial uses for water: brushing teeth; washing hands; showering; flushing toilets; washing dishes; laundry; other miscellaneous household cleaning; a means of hydration for animals.

CHAPTER XII
BREAD, PASTRIES AND PASTA

<u>Psalm 72:16; 104:14-15</u>: *May there be plenty of grain in the land; may it wave on the tops of the mountains. May its crops be like Lebanon.* <u>104:14-15</u>: *He causes grass to grow for the livestock and provides crops for man to cultivate, producing food from the earth, wine that makes man's heart glad—making his face shine with oil and bread that sustains man's heart.* (HCS)

<u>Hebrews 6:7</u>: *For ground that has drunk the rain that has often fallen on it and that produces vegetation useful to those it is cultivated for receives a blessing from God.* (HCS)

<u>Genesis 42:1-2</u>: *When Jacob learned that there was grain in Egypt, he said to his sons, "Why do you keep looking at each other? Listen," he went on, "I have heard there is grain in Egypt. Go down there and buy some for us <u>so that we will live and not die</u>."* (HCS) (emphasis added)

Bread is in the category of being sacred. Grain was highly respected and was the staple food that provided Life. Notice Jacob did not send his sons out to hunt for an animal to kill and bring home as food during the famine? I'm sure there were sheep or goats that could have been had as a meal. Genesis 42 tells us how important grain (that contains gluten proteins) is for the Life of the body, the Spirit Life.

Bread produces the best benefit to the body when it is eaten alone for breakfast and after sunset. Drinking water when eating bread will interrupt the process necessary for cell building and could cause fermentation issues. Avoid honey with any yeast bread, at least until there is time for testing to be done on this combination. In ancient times bread was the meal not an accompaniment to a meal. Bread goes well with herb toppings, olive oil, honey (on unleavened bread) or maple syrup. Organic fruit juice sweetened concord grape jelly is acceptable as an accompaniment for bread.

Gluten proteins, with the help of amino acids, produce hydrogen and is important for the development of healthy new cells in the body. I cannot stress this enough, proteins from a source other than glutens lack the ability or have a hindered ability to produce the necessary hydrogen for the development of healthy cells. The chemical explanation for the conversion and hydrogen development process is far too complex for me to attempt to insert into this paragraph but it is easily found on science websites. An easy explanation appears to be that more effort is required by the body to convert proteins that come from sources other than gluten to hydrogen for the development of cells. While other protein sources may cause the development of new cells the question is whether those cells are what they need to be for maintaining a healthy life. Note: Wheat and barley are mentioned most often in Scripture. There are other grains that contain gluten proteins, but I do not know if they would provide the same result as wheat. Personally, I stick with organic wheat flours.

To resurrect the origin of Hebrew DNA, gluten proteins from grain must be consumed. The Heavenly Gases must be in specific proportions for the body to produce the proper hydration and electrical activity to have the ability to eliminate harmful particles the body comes in contact with on a daily basis. A root to gluten intolerance is the lack of Love Exchange explained in Living by the Light of the Moon published by Harvest of Healing, 2024. We must be positioned to receive, process and return to the environment various Heavenly Gases. If that process is hindered or a particular gas is missing or out of balance, gluten protein digestion will be hindered.

Could cells that develop from meat or other proteins outside of gluten be a root cause of disease(s). A quick google search revealed the following protein related diseases:

Type 2 Diabetes
Inherited cataracts
Some forms of Atherosclerosis
Hemodialysis related disorders
Short-chain Amyloidosis
Creutzfeldt-Jakob Disease
Prion diseases
Alzheimer's Disease
Parkinson's Disease
Multiple System Atrophy, and more.

Breads should be made fresh daily, preferably without yeast and certainly without yeast during the first seven to eight days of Waning Moon phase. It is challenging to have the time to make fresh bread each day so freezing fresh bread the day it is made is acceptable. I make a batch of fresh tortillas and freeze the excess after they have cooled. Tortillas are quick to thaw and easy to reheat.

Two previous publications, Home-Made Answers for Cancer and Life Altering Disease by Harvest of Healing, LLC, published 2024; and Living by the Light of the Moon by Harvest of Healing, LLC, published 2024, cover the topic of yeast in detail. Yeast is not to be eaten during Waning Moon and afflicts the blood in ways yet to be discovered by science.

Shaping the Bread

The shape of bread appears to be just as powerful as symbols can be. Breads should not have holes in the center, nor should they be braided. The gluten in the bread provides necessary particles and gases for the production of healthy cells and a healthy cell does not contain holes, nor are they twisted. Loaf breads are preferred over dinner rolls and breads are to be broken, not cut with a knife.

Breads should not be accompanied by fruit, including raisins. This eliminates pies and all fruit filled pastries, fruit scones, and so forth. The combination of various chemicals and gases that develop when breads and fruits are combined will eventually disrupt the stomach. The first thing that comes to mind is elevated stomach acid and the lining of the stomach suffering. Fluid production and elimination become disrupted and will eventually result in sluggish lymph activity. The only fruit that is to accompany bread is concord grape.

Dry bread such as croutons or stuffing have no gluten value and are of no benefit to the blood. Something crunchy can be enjoyable at times but we also need what we eat to be a benefit.

Baking powder is permissible, but overuse can result in lack of tone or strength in the body. Imagine how biscuits rise in response to baking powder, this too will eventually manifest in the form of a large derriere through the generational lines. Weekly practice of the True Sabbath Meditation helps keep internal housekeeping in check and minor incidents from baking powder erased.

Pastries and Desserts

Who doesn't love pastries? I used to eat cake and cookies every day, until my health took a stroll through hell and back! Pastries and desserts are an easy way to step into gluttony. Cakes (cookies and fruit breads) contain eggs and eggs disrupt the function of the crown meridian (think of the cone on the top of the chicken's head). Mixing numerous ingredients into a "consumable" product causes confusion in the body, not to mention the chemical reactions that erupt. The body was not designed to process numerous food messages at one time.

Cheesecakes with fruit or without fruit, flavored cheesecakes, including pumpkin cause interruption in the intuition (spiritual hearing) and interferes with the energetic signaling of B and C vitamins. Consider a static electricity component as a possible cause for this interruption. Certainly, something is altering the blood. No doubt, cheesecake can leave a stain in the body that develops genetics for large statured individuals.

Layered pastries with creams and/or puddings and dusting of cinnamon create imbalance, or an instability. Here you must think of how moveable those puddings and creams are even though held within the structure of a cake or pastry. This would be like erecting a set of Legos but using Jello blocks as part of the structure. This instability would manifest as an inability to remain stable on your feet, a literal instability in standing or walking. It can also manifest as not being reliable.

Cookie dough, cake batter, frosting, including that found on those frosted and sprinkled donuts interfere with the electrodes. Decorated or frosted cakes and pastries cause a strain on the chest/breast area producing pain in the region of the shoulders. Celebrations with cake can result in callous on the toes, which is not only inconvenient but uncomfortable.

Ice cream (chocolate/vanilla in particular) interferes with the electrical conduction, influencing the right side of the brain. Hair can become altered and shoulder issues develop. Ice cream is guilty of genetic imprints for snotty noses in infants and children.

Pasta

Dried shelf-stable pasta has no gluten protein value. For gluten proteins to be involved in development of new healthy cells, it must be freshly activated. Dried pasta causes an energetic response that interferes with the meridians to the heart and conflicts with stomach energies. The meridians from the region of the heart down the left arm will deliver mild shocking sensations. These are not pleasant to experience. Avoid Angel Hair Pasta and Spaghetti, any long straight strands of noodle. Could the straight strands of noodles have an influence on the helix of DNA?

Pasta made fresh is a benefit in the gluten proteins department, a nice alternative to an unleavened bread as long as you do not add tomato sauce to your pasta! Keep the pasta simple, adding olive oil, a vegan pesto or fresh cut herbs. Sunflower seeds and artichoke hearts are good in pasta dishes as well. When consuming pasta, do so during a Waxing Moon phase.

<u>Deuteronomy 11:10-11</u>: The land you are entering to take over is not like the land of Egypt, from which you have come, where you planted your seed and irrigated it by foot as in a vegetable garden. But the land you are crossing the Jordan to take possession of is a land of mountains and valleys that drinks rain from heaven. (NIV)

To make the turnaround, people must leave the ways of the world, the "land of Egypt," and this includes the consumption of large servings of vegetables. The name Jordan means "down under," a reference to the internal river of the physical body, which involves the plasma. The body will experience mountains/ups and valleys/downs during the time of transition from the worldly lifestyle, but the source of life comes from the rain (Heavens) that falls on the wheat that is turned into flour, then into bread, and then eaten. This attribute of Heaven that comes through the grain is called hydrogen. The Heavenly Gases, which involves the hydrogen from gluten proteins, produces hydration in the body and will result in a disease-free life. This analysis should raise concerns for applying chemicals to crops and altering grain into a genetically modified organism. It is no wonder so many people have challenges with eating such grains! Yet another set of circumstances that developed deadly contamination all in the name of money! How to connect to this Heaven hydration is discussed in Living by the Light of the Moon by Harvest of Healing, LLC, published 2024.

CHAPTER XIII
CANDY AND CHOCOLATE

I don't know of many people that eat candy these days so I will not spend much time and effort on this subject. After learning what I have about food and its influence on the blood, my opinion rests that candy should be outlawed. It is more than obvious that flavored, coated, colored, cooked and molded sugar is not a beneficial product for consumption by anyone. These ingredients remove the light from the plasma. It is no wonder children spend the months of November and December not feeling well after bags of Halloween candies are consumed. Yet another American tradition that needs remodeling.

A few notes on this subject: peppermint will eventually contribute to the hair falling out; candy can result in womanly traits being removed; candy interrupts sleep stages (1, 2, 3 and REM); and chewing gum causes a delay or misfire in the electrical system in the body. We can conclude that anything classified as candy will likely land you at the medical facility. Could the sugar spike be the genetic root of diabetes? If so, we can throw alcoholism in as a guilty party for diabetes as well.

Chocolate is a challenging subject. Who doesn't like to have chocolate now and then? There is a chemical response triggered by the cocoa bean that results in irritation of viral infections, loose bowels, influence directed at the shape of the nose and mouth when accompanied by cherries, and health of the skin.

Aggressive personalities can development over time through generations as a result of consumption of the cocoa bean. That cup of hot chocolate that sounds and feels so good when you are under the weather or have a bit of a chill may not be such a great idea. These irritations and alterations can also reach to the depths of genetic imprints. For example, if you inherited a herpes virus, the virus is unreachable by medical intervention but on occasion produces a cold sore or inflammation response, cocoa is a trigger for resurrecting the dead (genetic imprints) and symptoms of cold sores or inflammation are experienced.

Chocolate, dark, milk or white, can irritate bowel regularity and over time influence the function of the pancreas resulting in pancreatic cancers, emotional disturbances and diabetes. Chocolate will cause dampness in the body, interrupts intuition reception, and causes the brain electricity to periodically switch places. We must not leave out the all-too-common problem as people age, the random hair growth in the nose and around the region of the mouth. Chocolate, when accompanied by dairy can alter chromosomes. Chocolate will interfere with normal sexual response or attraction. A type of roadblock is erected in general functions in this arena. There are documented studies of how cocoa powder has an influence on the brain but here again, those studies only report on any positive result not taking into consideration the influences contrary to their reported findings or how cocoa responds when mixed with other foods. Chocolate mixed with peanut butter, in cookies or pastries attacks the electrical aspects in the plasma causing the plasma to lose its light/electricity. Chocolate with milk and eggs has an influence on body stature. Chocolate can result in being sleepy during the day and a tickling or tingling sensation on the bottoms of the feet. This could be an answer for neuropathy.

CHAPTER XIV
CONDIMENTS

Condiments contain vinegar which is a "no-no" for those who are attempting to achieve a Hebrew DNA status. No doubt, condiments are an avenue to increased salt and sugar intake. The sugars, likely corn and cane are used, disrupt the Hebrew DNA. Both corn and cane sugar require a fair amount of water during their growth and when consumed leave a "need for water" signal in the cells of the body. Ingredients for condiments are processed and with those various processing steps, the value of any fruit or vegetable is lost. The most beneficial route is to set all condiments aside.

Pickles and other vinegar preserved foods have an energetic connection to brain cancer. This could rest in the use of vinegar, or it could be a combination of vinegar with any fermentation process. Vinegar afflicts the life in the cells. I won't even use vinegar as a cleaning agent. Things that go to the gut will gravitate to the brain like bubbles that rise to the top of a glass.

Salt should be used sparingly when going through any cleansing protocol described in Home-Made Answers for Cancer and Life Altering Disease by Harvest of Healing, published 2024. Table salt will cause the heart to flutter, giving good reason to eliminate the salty chips.

CHAPTER XV
DAIRY

Manipulated reports and studies have driven the sale of milk and dairy products hard. Dairy has some aggressive components, one being stress signals from the cow. There is good reason why babies can be challenged with consuming cow milk.

<u>Deuteronomy 32:13-14</u>: He made him ride on the heights of the land and eat the produce of the field. He nourished him with honey from the rock and oil from flint-like rock, cream from the herd, and milk from the flock, with fat of lambs, rams of Bashan and goats, with the choicest of grains of wheat; you drank from the finest of grapes. (HCS)

Bashan is the northernmost region of Transjordan, modern day Syria. Jordan means "down under" (inside the body) and being in the northern region could reference the head or brain. Other references include underworld and serpent or dragon as meanings of Bashan. If you must supplement milk, use sheep or goat milk.

Any consumed dairy that is from the herd must contain the cream to avoid changes in the chemical structure. The chemical structure of the milk is altered and becomes a negative influence on the inner workings of the body when it becomes separated from the fat. Standardized processing in general has its downfalls as well and contributes to the alterations the chemical makeup of the milk incurs. Cream or curd is to come from the herd (cattle); milk is to come from the flock (sheep or goats.) While consumption of these products is permissible, it is advised to limit the amount and frequency of such consumption. I discontinued use of milk and consumption of cheese made from cow's milk. I use cream to make pancake and biscuit batter. I do eat sheep cheese on occasion, being sure to avoid the first seven to eight days of Waning Moon. Cheese will cause you to look older than you are. Just like cheese is aged, you will age as well. Cheese can also upset the stomach. I do not drink wine, but I would assume this same aging message is present in wine and other aged foods.

Overuse of butter will produce pain and discomfort in the neck and upper spine, shoulder region. While butter goes well with warm bread, it should only be eaten at times of fine dining.

Packaging milk in plastic containers increases the risk of bacterial growth. Plastic will heat and cool quicker than glass and the chemicals used to produce the plastic can have an influence on the chemical structure of the milk. Newly discovered genetic mutation GRIN is the identification of calcium penetrating the cell walls causing a calcium overload within the cell. The calcium inside the cell would ultimately snuff out the electrical activity within the cell. Could calcium from cow's milk be the root cause of this genetic mutation? Could there be an improper form of calcium because of the removal of cream and the processing of the milk?

Some additional discoveries produced by milk and milk products that may be of interest:

A Jewish diet often includes avoiding meat with dairy, and for good reason. This combination can result in an emotion of grief and the loss of the hair on the head. Add meat to dairy and you could end up with development of red patches on the skin, eczema or birth marks.

Whipped cream influences the meridians in the neck and shoulder region and will result in pain;

Cottage cheese is foreign to the Hebrew DNA; cottage cheese contains signals that attract to the brain and can result in disruption in brain function and health. This is a good example of how foods that are similar in appearance to an organ, system or gland in the body will have an influence on the look-alike location.

Yogurt can cause a depleted spiritual (electrical) function in the body, cause stomach function issues that can eventually graduate to chronic stomach malfunction and genetic imprints that result in future generations having stomach issues. Yogurt also promotes memory issues.

Frozen foods cause the shoulders to lean forward and the head to bow creating in a burden on the body. This would include milk shakes and other frosty or frozen drinks. These frozen food items cool down the interior of the body too quickly and the cells respond to that temperature change. Standing or sitting in cold water is not a good idea if you plan to have healthy cells and cell signals.

Being petite was lost by consuming chocolate milk shakes and egg soufflés and many vaginal yeast infections are rooted in and irritated by milk products. There is a chocolate, milk, egg combination issue at play.

I'm not aware of any acceptable milk substitutions, emphasis on acceptable. There are many milk alternatives, but they are not always what the body can use and at this point the level of harm they may inflict is unknown. Baby formulas and canned milks like Pet Milk have influence on the genitalia, particularly the penis.

CHAPTER XVI
FRUIT

There is no fresh fruit that will result in a cancer. Properly ripened and washed fruit actually helps clear out some of the bad vibes that lead to sickness or disease.

Most fruits are to be eaten separately from other food categories. The only exception to this that I have found is the concord grape. Concord grape benefits the crown of the head and helps clean the blood. Fruits mixed with other food categories can result in a genetic imprint for the "anything goes" lifestyle and disrupt the elemental gas Argon. Fresh fruits in general are good for the blood.

Fruits should be fresh, not canned or frozen, when possible. Rinse all fruits under running water and eliminate peelings. Sadly, dairy does not mix well with fruits. So much for the whipped cream on berries. On occasion, I enjoy sunflower seed butter with fruit. Add a little honey or maple syrup to the sunflower seed butter on days of the Waxing Moon phase and it is a nice, sweet treat.

Bananas should be eaten while they are slightly green. Over-soft bananas trigger a histamine response and will evolve into a Mast Cell Activation Syndrome response. Bananas help reduce internal heat issues and are a good source of Chromium, a nutrient beneficial to the Star Dust and mRNA process. Eating banana helps correct pH balance.

Grapefruit is harmful to Hebrew blood chemistry. Packing and shipping of citrus fruits could have an impact that results in the lack of harmony with Hebrew DNA. Citrus fruits have an elevated water content and water can record vibrations it comes in contact with. Lemon in small qualities has not raised any flags; I cannot say the same for lime. Lime appears to have the intoxicating type influence on the blood.

Nothing is better than a nice cool, crisp slice of watermelon on a summer day, except that it can create too much fluid in the body, drowning out the light in the plasma. Limit watermelon or eliminate it, the fluid accumulates in the body and can result in memory issues or dementia, fluid accumulation on the spine or in the brain. Melons can trigger inflammation responses. Eat tropical fruits (mango, guava, papaya) in place of watermelon.

Berries are a good option. Strawberries are good for making into a tea, not necessarily for eating. Strawberries can host mold in those tiny little seeds that cover its exterior, not always easily seen. Strawberries are a high histamine food and can trigger an inflammation response. Extra caution should be used when washing and preparing berries. They spoil easily and molds can hide in their creases. If raspberries have tiny black spots in them, do not eat them. The black spots are mold sprouts. Blackberries and blueberries: if there is any white speck on them, toss them out. Any berry that is not firm should be tossed out. These molds can initiate existing infections, including those that are in genetic imprints.

Berries will correct a crook in the bridge of the nose that initiated from eating roast beef and fresh raspberries help clear out cancers. Hebrew DNA loves peaches without the skins. Go easy on consuming pomegranate seeds. Like many things, too much of a good thing can lead to a bad thing. Cherries are messengers for the body, helping clear out signals from baked goods. Cranberries are not good for toddlers and should be avoided during the month of December, the exact time everyone likes to eat them. I'm not sure why December is crucial. There would be a connection to the Heavenly Gases and specific activity taking place in the realms we do not see during that time of the year. Use caution with tomatoes. If any of the seeds appear black, the tomato has tiny bugs on the inside. Bacteria will grow quickly on damp fruit so wash any fruit just prior to eating.

Fruit breads spoil rapidly and should be avoided due to the mixing of grain with fruit. There is a fermentation or chemical reaction issue here.

CHAPTER XVII
GRAINS

If not stored with proper ventilation, grains can easily become infested with mold. Store grains in glass containers or cloth sacks and place them in the refrigerator. It is best to avoid boxed grain mixes just for the simple fact that the less manufacturing, the more natural the product.

Corn, as it is known in the U.S., the yellow or white kernel grain that often is fed to cows, causes sluggish lymph. Something about corn has never agreed with my body and that comes from allergy and energy testing, not from a physical symptom. The lack of a noticeable symptom tells me something about the entire makeup of corn conflicts with the Heavenly Gases inside the body. Corn has become known for its ability to produce ethanol, a colorless, volatile, flammable liquid. Herein could be the answers to the issues with consumption of corn. A fermentation process is used in the production of ethanol and that same fermentation process could take place inside the body with the measure of a sugar and body heat. I have learned that consumption of corn can contribute to susceptibility to sunburns.

Rolled oats lose their vitality through processing and mucous can be a result interfering with the microbiome in the gut. Organic steel cut oats are an acceptable option with the note that oats do not contain gluten proteins. When I eat oats, which is rarely, I become hungry within an hour or two. I can eat an equal amount of cream of wheat and not experience hunger pangs for several hours. The gluten proteins provide the necessary components for keeping the body satisfied.

Wheat lacks Lysine. Lysine is/was used to treat herpes and may be a form of relief for those who experience genetically inherited cold sores. Sources of Lysine: potato, pears, tomato, leek, beets, apricots, avocados, mango, green/red bell pepper, pumpkin seeds, macadamia nuts. Lysine is also available in supplement form.

Gluten is activated when combined with liquid. The structure of gluten proteins change when the baked good is left overnight on the countertop, causing it to become less beneficial in the development of healthy cells. To stop the gluten proteins from changing structure, bread products must be frozen the day they are made.

As stated before, there is nothing beneficial about dried bread, croutons, pita chips and so forth, other than the satisfaction of the crunch. Fresh breads are the only source of beneficial gluten proteins.

CHAPTER XVIII
MEATS AND PROTEINS

Meat can be eaten when specific guidelines are in place. During the Waxing Moon phase, meat is to be eaten at twilight only, eaten with haste while dressed in appropriate attire. (Exodus 16). Meats may be eaten Tuesday through Thursday, avoiding the Friday through Sunday True Sabbath Meditation days. Blood infection can develop from eating meat outside of the guidelines set forth in Scripture, and eating meat that has been frozen will eventually result in the body becoming stiff with the potential for head and shoulders to bow forward. The body will copy the message you give it. It is likely not age that is causing those stiff legs or stiff back but years of consuming meat that had previously been frozen.

As a continuation of the information previously shared on the structure of cells, could protein related disease, "CF" (cystic fibrosis) for example, genes be a result of consuming an incorrect protein? CF is present in Hispanics at a greater rate than other ethnicities. Reaching back 50 to 100 years, could consumption of bean protein be a root cause of CF genetics? I doubt the structure of various proteins has been given much consideration by research. Proteins seem to have been put into the same collective pot when one source of protein can differ greatly from another source of protein. To put them all in the same collective pot would be like saying all flowers are the same. Considering the various genetic disruptions and cell malformations, it would only make sense that there is a component within the structure of the protein and the source it originates from that is influencing the negative return being witnessed. It is possible that sources of protein outside of those in gluten can be consumed to a certain degree but what if the consumption of alternate proteins tips the scales and produces a result in malformed or unhealthy cells? Could too much of a foreign protein contribute to mental health issues such as schizophrenia? What about high cholesterol being a result of the consumption of incorrect proteins versus the common belief that it comes from eating red meat.

One final notation here, it is advisable to avoid smoked meats. There have been varying reports on the influence of smoked meats and I have learned of a few cases where those who ate smoked meats several times per week produced serious health issues. Hydrocarbons present in the smoked meat is a factor and possible guilty contributor to stomach cancer, cardiovascular disease and high blood pressure.

Fish and Seafood

Signals from foods can do odd things to the human body and fish is no exception. Fried and/or breaded fish can result in a genetic imprint that causes a desire to live as/be the opposite sex. Consumption of fish can result in deviated septum, a large head (consumption of tuna), flaky or oily skin, birthing large babies and a tearing away or flaking of the flesh or tissues, similar to how the flesh of the fish flakes apart. With this much influence on the head, what might this fish diet be doing to the crown meridian? One must assume this is all dependent upon the type of fish being consumed and accumulation of signals coming through generations.

Consuming soups or other food dishes made with fish heads can result in genetic imprints for hyperactivity and brain disturbance, including cancer. Raw fish has never seemed appetizing to me and if you have ever watched the television show Monsters Inside Me you might change any liking you have for raw fish. Worms and parasites come to mind when thinking of raw fish or fish eggs. Sushi is loaded with salt and we've all heard how harmful excess salt is to the body. Fish can contain high levels of metal such as silver. Shellfish is a trigger for inflammatory response and should be an obvious message that the physical body is not designed to tolerate it. I have serious doubts that sardines are all they are reported to be either. Overall, fish has influence on the cell communication activity in the body. As the old saying goes, "you are what you eat."

Poultry

When you look at a chicken, or other foul, what do you see? What you see is basically what you will get in terms of the body responding to the messages received from the chicken, or even the eggs. When live cells from animals are consumed, a signal is received by the body that will mirror certain aspects of those cells. For example, if you eat an egg, you are eating cells that eventually, through the generational line, will produce a nose of greater size or extension than one who has no consumption of eggs in their genetic panel. Think of the rooster's head with a beak and red comb on top of the head, these are the physical features eggs will change in humans. Have you ever heard the phrase: "Look at the beak that guy/gal has!" This large beak/nose is evidence of the consumption of eggs somewhere along the ancestral line. Eggs will also shift the location of the crown on top of the head. This can be troublesome when it comes to receiving electrical transmissions from the Heavens. When the crown chakra/wheel is shifted it will eventually result in a spiritual (electrical) death. Eggs are a contributor to strokes. Eggs contain undetected bacteria, and this would trigger the interior temperature of the body to become elevated. Too much heat can cause condensation and result in bacteria growth. Eating chicken irritates the plasma in the body and production of phlegm will increase, a sign of inflammation or infection. We may be able to chalk some of this heat issue up to many years of eating fried chicken. Who doesn't like fried chicken!? Heating foods to an extreme by dipping them into hot grease does not produce a safe to consume dinner. The level of heat the food incurs influences the temperature in the body and that elevated temperature can become recorded in the cells, depending on how often you eat fried foods. Cooking meats on a gas or charcoal grill can contribute to the excess heat issue as well. Aside from all this, eating chicken or eggs can leave a person feeling hungry. There are no gluten proteins in chicken or eggs and no gluten proteins means becoming hungry due to the lack of necessary hydrogen.

If I haven't convinced you of the potential dangers of eggs, I'll throw in that separating an egg yolk from an egg white creates an incomplete protein.

Job 6:6: Is bland food eaten without salt? Is there flavor in an egg white? I refuse to touch them; they are like contaminated food. (CSB)

According to Job, salt must accompany an egg white. I think we can safely assume it is the minerals in a good quality salt that are necessary for an egg white to be used by the body. Proteins are necessary for the building of new cells so what are we doing to our cells if we are consuming an incomplete protein? At this point, I'm not convinced that eggs are an acceptable protein for the development of healthy human cells.

Avoid using olive oil on chicken, energetically they do not mix well that could result in a toxic issue. This needs professional eyes to determine exactly what takes place when chicken and olive oil and put together. Olive oil is best if left for baking and dipping breads. Avoid cooking chicken in a microwave. For that matter, avoid cooking anything in a microwave.

As we've all heard, turkey, or any other meat, should never be thawed on the countertop. Wrapping meats in plastics is another avenue for the blood to become damaged. Turkey contributes to brain cancers and initiates genetics for a lack of motivation and laziness.

Red Meats/Beef

Red meat is a topic all its own. To eat the blood of another living creature that carries a level of its own energies received from the cosmos should be evidence enough that consumption should be limited. God owns the cattle on a thousand hills telling us that the energies from the cosmos are received by the blood of cattle, and likely all other red meats. Does this cosmic energy make the meat safe for consumption? No. The level of or types of elemental gases received or used by a cow can be quite different than those safe for the human body.

Meat is to be cut in a specific manner, and I can only assume this is to avoid any potential negative influence being transferred during consumption. Obesity has a connection to meat being incorrectly cut. Bite size meats, particularly when prepared in Asian food dishes do not break down properly in the digestive system and become bound up, hard to process.

If you eat a steak and the location that steak was cut from had incurred an injury during the life of the cow, that injury signal no matter how long ago it took place, can come through to the human consuming the steak. Blood has a language and an ability to communicate. Steak also has a connection to sweating and prostate cancer.

Consuming roast beef will cause a genetic imprint for a crook in the bridge of the nose; cuts of beef from the shoulder region can result in issues with the shoulders. Brisket cuts contribute to white/gray matter in the brain and can afflict breast tissue leading to breast cancers. Prime Rib afflicts the memory replay and interferes with ability to receive signals from the cosmos. Prime rib can hinder the male ability to produce offspring, and that issue will reflect in a pair of genes.

Ever notice the cold, wet, runny nose of a cow? Eating beef can result in lung infections from too much fluid accumulation in the chest cavity. This list could just keep going.... The good news is fresh berries will help the body eliminate some of these harmful messages. Wearing natural fabrics and colors also have their benefits in this area.

Raw hamburger, not fully cooked or improperly thawed prior to cooking, can easily spoil and ends up causing an elevated heat issue in the body. Hamburger assaults the brain, particularly in women. Think about it, ground beef has a similar appearance to that of a brain.

Hamburger can cause the body to resist conceiving, promotes aging, causes an intoxicating effect on the blood, and when cheese is added to the mix, dry lips can be a result. Issues with conception may manifest in the form of a person not being interested in children or an actual medical issue involving the reproductive organs, hormones or glands. Previously addressed is the signal given off by consumption of multiple food groups producing an anything goes mindset. This rings true with consuming hamburger as well. Leviticus 4 talks about all the fat being removed and the fat being burned on the altar. Hamburger contains fat at different percentages and the reference to burning the fat indicates heat being involved. Next time you think that hamburger from your favorite restaurant or café sounds good, it might be wise to reconsider the thought.

Bones and bone broths influence the plasma, resulting in an interruption in the Living Water process necessary for eliminating harmful vibrations or debris. Cooking meat with the bones will eventually result in issues with the spine and excess water production or accumulation in the body. The subject of bones brings up gelatins. Gelatin interrupts the healthy protein that is necessary for the growth of new cells. These new cells need to have the ability to receive the electrical signal from the cosmos and have an ability to carry that signal. A new cell can develop but how healthy or how well charged is that cell? This is an area for science to unravel.

On the same level of interest as the fish for me, organ meats, intestinal linings, tongues, and other parts should be ruled uneatable. I can only imagine the signals being sent to the blood when you eat entrails of an animal. The unseen elements that are contained in these animal parts can easily cause an imbalance in our Heavenly

Gases. Continually consuming these things can result in a part of the human body shutting down, something will be thrown out of balance.

Leviticus 4:11: But the hide of the bull and all its flesh, with its head and shanks, and its entrails and dung – all the rest of the bull he must bring to a ceremonially clean place outside the camp to the ash heap, and must burn it on a wood fire. It is to be burned at the ash heap. 9:14-16: Then he brought the bull near for the sin offering, and Aaron and his sons laid their hands on the head of the bull for the sin offering. Then Moses slaughtered it, took the blood, and applied it with his finger to the horns of the altar on all sides, purifying the altar. He poured out the blood at the base of the altar and consecrated it so that atonement can be made on it. Moses took all the fat that was on the entrails, the fatty lobe of the liver and the two kidneys with their fat, and he burned them on the alter. (HCS)

The verses in Leviticus are talking about the influence on the blood from consumption of beef. Fire is purification by heat meaning the heat can become elevated in the body to remove the vibrations (sin) that are now in your blood. Notice the mention of the liver and the kidneys? Could consuming beef be a contributor to health disruption in the liver and kidneys? Again, the number of years of consumption and whether ancestors were also meat eaters would need to be taken into consideration. There are numerous ways something like beef consumption resulting in liver or kidney health issues could be configured.

Repetitive consumption of grilled meats and meats cooked on open fires result in a need for vision correction, and a desire for wearing the hair short or shaving the head. Consumption of meat from a black cow can result in nose/mouth alterations or allergies. I have not observed cattle eating outside of what is put in a trough or from a distance as they graze in the grass, but it would make sense that a cow that eats weeds and grass can have an impact on allergy issues one might have in response to eating meat. It will be interesting to discover what the consumable difference is in the meat from a red cow versus the meat from a black cow.

Jerky meats and sausage have a connection to knee issues and canned meats are better fed to your pets.

Any fermented food or drink will add complications to what is being eaten. Consuming alcohol when eating red meats can result in genetics for muscle pain and injuries in response to the electrical pathways being influenced.

Pork

Pork interferes with the electrical balance (yin/yang) of the body. Bacon causes things in the body to stick together and elevates the internal heat in the electrical receptivity component in the body. Could this sticking together component reflect in the DNA strands or brain tissue?

Soups with pork are connected to keeping secrets, and consuming pork intestines can result in relationship conflicts. Hotdogs are a processed product and are very low on the list for consumable products in my opinion. Do we really know what all is mashed and packed and formed into a hot dog? To add to the mystery, what sort of signal is the body receiving when we consume a food that is similar in shape to a male body part? May as well throw bratwurst in this group while we are at it. Words must be withheld from paper for what this has evolved into. It can take up to two years to eliminate the damage done inside the body from the consumption of hot dogs. The fact that legions of demons were cast into pigs by Jesus gives rise to the question of whether eating pork results in a multitude of damaging health issues. The demoniac the demons were cast out of had a host of health-related issues.

<u>Luke 8:26-33</u>: *Then they sailed to the region of the Gerasenes, which is opposite Galilee* (wheel, whirlwind, skull or head; cyclical action). *When he got out on land, a demon-possessed man from the town met him. For a long time he had worn no clothes and did not stay in a house but in the tombs. When he saw Jesus, he cried out, fell down before him, and said in a loud voice, "What do you have to do with me, Jesus, Son of the Most High God? I beg you, don't torment me!" For he had commanded the unclean spirt to come out of the man. Many times it had seized him, and though he was guarded, bound by chains and shackles, he would snap the restraints and be driven by the demon into desert places. "What is your name?" Jesus asked him. "Legion," he said, because many demons had entered him. And they begged him not to banish them to the abyss. A large herd of pigs was there, feeding on the hillside. The demons begged him to permit them to enter the pigs and he gave them permission. The demons came out of the man and entered the pigs, and the herd rushed down the steep bank into the lake and drowned.* (CSB) (description added)

Highlights in the verses: adversity coming to the region of the skull; legions reflect many health issues. Pigs were the recipients of the "demons" that result in multiple health issues causing one to lose their spiritual covering and live a physical life amongst death (of the Soul). The pigs moving over the bank into the lake is reflective of things becoming out of control with respect to the plasma.

For those of you who are now disgusted with this insight, drink concord grape juice to help the body remove the signals it has received from meats.

CHAPTER XIX
NUTS, SEEDS AND OILS

I've selected to stay clear of cashews, and pistachios can cause a histamine or inflammation response. I purchase raw, unsalted, shelled nuts, usually almonds. I rinse and soak the almonds over night, strain, rinse and remove the skins the next morning. Spread them onto a parchment paper lined baking sheet and dry in the convection oven at 225 degrees Fahrenheit. When they begin to turn light tan in color, they are dry. To sprout sunflower seeds I use the same process. I do not sprout pumpkin seeds, I rinse them then dry them in the convection oven. Rinse pecans or walnuts and dry in a convection oven and store in a glass container. After eating nuts, rinse the mouth to remove nut particles from the teeth and gums.

We all know it is advised to not give children peanuts due to the choking hazard. What is not as well reported is the histamine trigger that peanuts are guilty of. What may also be news is that children should not eat pecans. It is necessary to be very cautious with pecans because of the deep creases and folds in the structure of the nut, and walnuts have this to a degree as well, molds can take up residence in the creases. Personally, I limit the consumption of nuts. Eyelids that have a pink appearance may be a sign of over-consumption of nuts.

Olive oil is most beneficial; sunflower seed oil is acceptable. Avocado oil can trigger a histamine response so use it sparingly or with caution. No coconut oil, corn oil, vegetable oil or vegetable shortening. Avoid using olive oil on poultry and in salads. Olive oil is most beneficial when it accompanies fresh bread.

CHAPTER XX
SNACKS AND MISCELLANEOUS

Colored sugars, such as flavored and colored suckers or other hard candies cause blemishes on the skin. Sugar cane influences the hair causing the hair to lose its natural wave or curl and the hair becomes straight and weak. The hair is one feature that attracts the Heavenly Gases, giving explanation as to why woman from many generations wore their locks long. Candy drains the Heavenly Gases from the body.

Certs mints contain Retsyn (copper gluconate, tetra sodium, pyrophosphate), sounds complicated and appears self-explanatory as to why they should be avoided.

Chewing gum influences the strength in the teeth and gums. There are nerves in the mouth that will send signals to the blood and excess chewing, sugar and, even worse, Xylitol influences the signals received by the blood and those signals eventually make their way to the brain. Juicy Fruit and fruit striped gums cause a need for the body to perform extra cleansing and could be said for all chewing gums to a certain degree. The fruit flavored chemicals added cause complications for the body to address, relative to the liver. Chewing gum can also result in bright red hair and a snotty nose. It takes several months to erase the influence of chewing gum. Here again, do we really know what we are putting in our mouth?

Peanut butter influences the complexion. The oil in peanut butters will send a signal for the skin to be oily.

When eating sunflower seeds be sure to remove the shells, rinse under running water and dry them in a convection oven. Tiny worms like to make their home in the sunflower seed shells.

Popcorn, popcorn balls and Cracker Jacks all have a signal that will influence testicles. Given the shape of a popped corn kernel, it is no surprise where the location of influence would be, in what manner of influence remains the question. I'll leave this for the men to figure out.

Crackers fall in the category of dry breads and have zero beneficial hydrogen influence.

No raisins, like all dried fruit there is no benefit left in them when it comes to the ability to produce a healthy chemical reaction.

Table salt with pepper, particularly on popcorn, causes a narrowing in the pathways of the meridians, and table salt, such as Morton's and other mined generic brands, can influence the sense of smell and cause the heart to flutter. No doubt salt has an influence on the electrical activity in the body. Leaving salt and pepper on the table during a meal will cause digestive upsets.

The vanilla bean is another troublesome bean. It smells good and adds nice flavor but this is another one that has raised red flags for me. It has been challenging to pinpoint why so an educated guess would be interference with the Heavenly Gases. Beans in general appear to interfere with the Heavenly Gases in the body.

As previously stated, coconut can irritate the lining of the stomach.

The female monthly cycle can be influenced by many of the commonly used spices, extracts and even snack foods. For an alternative to eating ground spices, see the recipe I use for Spice Water at www.HarvestofHealing.com.

The good news is, sleeping on pink sheets helps the body eliminate the influence of these sugary, colorful products.

CHAPTER XXI
SUPPLEMENTS

Vitamin D supplements can cause puffy eyes.

Chromium is necessary to protect the mRNA signals for duplication of healthy DNA. Optional forms of intake are a chromium supplement; eat banana, apple, whole grain/wheat, drink grape juice.

Supplements are available in all forms for nutritional and medicinal purposes. Consult a professional for guidance with supplementation. As a person develops increased Hebrew DNA, the need for herbs or nutritional supplements will diminish.

CHAPTER XXII
VEGETABLES

<u>Romans 14:2</u>: *One person believes he may eat anything, but one who is weak eats only vegetables.* (HCS)

Vegetables may not be all they've been reported to be. Many foods, including vegetables and herbs are intended for medicinal purposes, not everyday consumption. Some foods are to be eaten at specific times of the day, which may in part have something to do with the position of the sun and its energy influence on the body's ability to process what has been consumed. The reference to being weak having a connection to vegetable consumption tells me vegetables influence the health of the chromosomes to a certain degree. When chromosomes are altered, or not healthy, the body will display a level of weakness.

Some popular food items, particularly in the vegetable category, are boasted for assisting the function of one part of the body or organ, yet causing harm to another, though the harmful aspects are never reported to the public. Simple things like garlic that is touted for its boost of the immune system and may be good in times of a cold or flu, but an overabundance of garlic can lead to brain cancer. This might make you think twice about eating that extra slice of garlic bread! It is a very potent herb. Broccoli is another vegetable that will deposit its essence in the brain. Vegetables that have pods like peas or green beans can harbor worms. Eggplant interferes with the Life in the plasma. Bell peppers contain too much water, as do many other peppers. Squash and zucchini are in this category as well.

Eating multiple vegetables at one time sends multiple messages to the body that it cannot always read. Similar to having too many pop-ups on a computer screen, there is too much activity going on at once. Too much of one nutrient could easily throw another nutrient out of balance and that's not even considering the gas activity that takes place. I experienced a time when I was adjusting my diet attempting to calm hyper inflammation responses and it seemed the more vegetables I ate the more irritated my body would become, including worsening of sleep interruptions.

The body is highly intelligent, if it is not abused and is maintained properly. Eating a large bowl of various greens with radishes, carrots, tomatoes, croutons or pumpkin seeds may sound good but the body may not be processing it all as we think it should due to overload of information. The body can develop a taste or craving for a particular vegetable in response to consistent exposure to a color. If you are consistently exposed to the color green, a vegetable of like color will be desired. Some may not need or have an ability to process the types or amounts of vegetables recommended by health professionals.

Vegetables can put a strain on the gallbladder and leafy greens were the guilty party for me. Leafy greens are praised for the nutrition and hydration abilities. Vegetables in general are challenging to digest and will disrupt the sleep when eaten in the evening. Vegetables may be better processed by the body when consumed without the accompaniment of another food group.

Items like pumpkins that are sometimes bedded amongst straw due to the cold nights, will be visited by rodents and birds and those visits may not always be visible. Leaving a pumpkin on the porch as a Fall decoration may be best.

Too many onions can irritate the breast tissue and, like garlic, results in the body producing an odor, not just bad breath. Onions also win an award for causing chronically chapped lips. French fries will influence the forebrain and too many carrots will influence the ears. This could be why rabbits are often pictured with carrots. Overall, I think it is safe to limit vegetables to special occasions or for medicinal purposes. Vegetables have their place but over consumption of vegetables can have its drawbacks. A homemade vegetable broth may be the best option.

No doubt, vegetable gardens contain many hidden critters or the residue they leave behind and may not be the best option for a clean diet. Lastly, if you eat vegetables during a Waning Moon phase, the gallbladder becomes disturbed.

CHAPTER XXIII
APPLIANCES

This will be something many will not want to hear. A dishwasher in operation will have an influence on the cells. The movement of water combined with heat has a form of impact on the cells in the body. How? Water in many ways is very influential to the body. It is as though the cells follow the lead of the water. A dishwasher recycles water within a contained space, basically sharing debris from one dish to another. If this is the signal that the cells in the body receive it could create situations where the body is recycling a portion of its fluids within rather than eliminating contamination. If you want clean dishes, cool water kills more bacteria than soap and warm or hot water. Hand wash your dishes and rinse them in cool water. Dishwashers will increase bacterial growth, harboring the contamination inside the dishwasher and then distributing it throughout the dishes during a cycle.

The color of appliances has an impact on the atmosphere and energy of our home. Appliances should be a cooling color helping to offset the heat generated by use of the appliance (cook stove, oven) or the electrical motor that keeps the appliance running. Stainless steel is a static electricity generator and should not be in a kitchen. The avocado green craze of the 1970s did its share of damage in kitchens and is not a beneficial choice of color for a kitchen. White is an option; aqua and lavender are cooling colors.

It should be obvious by now that microwaved foods or drinks are not beneficial. In fact, the microwaves deposited into the food can result in being tired. There is a commotion between the cells and the microwaves that do not get along.

Foods stored in the refrigerator will begin to accumulate condensation. This condensation contributes to hair loss. Toss out all foods with condensation build up. Refrigerators should be cleaned regularly by removing any expired or molded foods and wiping shelves with a damp cloth. All contents should be kept in an orderly fashion. This keeps the atmosphere inside the refrigerator in good conduct and the foods happy.

CHAPTER XXIV
EMOTIONS

Children, adults and everyone in between those two age groups have some level of emotional reaction to various things. The troubling issue is that the emotional flareups and uncontrollable fits seem to be on the rise. The book of Mark sheds a little light on what is taking place inside the body.

<u>Mark 7:17-23</u>: *When he went into the house away from the crowd, his disciples asked Him about the parable. He said to them, "Are you also as lacking in understanding? Don't you realize that nothing going into a person from the outside can defile him? For it doesn't go into his heart but into the stomach and is eliminated" (thus he declared all foods clean). And He said, "What comes out of a person is what defiles him. For from within, out of people's hearts come evil thoughts, sexual immoralities, thefts, murders, adulteries, greed, evil actions, deceit, self-indulgence, envy, slander, pride and foolishness. All these evil things come from within and defile a person."* (CSB)

These verses are clear, foods that are contrary to the Hebrew DNA will eventually lodge in the genetics and emotional chaos will result in some form or another. The attributes listed in the verses are a result of or remaining essence from foods. The food itself passes through the digestive system. The essence left behind will eventually burst forth in the form of an emotion.

Foods can result in various emotional disturbances and when stacked one on top of another a person can end up an emotional mess. Emotions are not solely a result of life experiences. A good consumable emotion example is chicken. Observe the chicken in its nervous fluttering about here and there. If enough chicken is eaten by a person or their ancestors and that person or ancestor does not have the proper True Sabbath Meditation practice for their Heavenly Housekeeping in place, eventually there will be descendants born with a tendency to be a Nervous Nellie.

Anger can be a result of continuous years of eating beef and a few other foods win this award as well. Have you witnessed the bull or cow stomp its front leg on the ground in a sort of pawing or digging motion? Again, this emotional development is not only by a single person's consumption of beef or any other food but likely the accumulation that comes through the genetics from the ancestors.

Jesus never really seemed to express a lot of emotion. Most often He is seen very focused and seemingly unmoved by what was taking place around Him. Two occasions come to mind where He displayed emotion: 1) when He turned the money changers tables over at the temple complex (Mark 11:15); 2) when beads of blood in the form of sweat were apparent on His brow (Luke 22:44).

If I could write a paper on the origin of emotions this is what it would contain: emotions develop in response to residue from various foods and/or consuming a collection of foods. One food may not propose an issue but when that food is consumed with another food of an opposing vibration or essence, it is possible to develop an emotion. The chemical reactions of various foods can clash and a harmful or opposing vibration is birthed. That family favorite casserole could be the root cause of the family feuds that develop over time. No doubt, some emotions are a result of a life experience, particularly when a person is exposed to a traumatic experience. Stuck or reoccurring emotions can also be a result of eating during times of elevated emotions.

This brief list of food related emotion is just what I have become aware of. Any food can become attached to an emotion if it is eaten at the time the emotion is active. This is one explanation why King David refused to eat or drink until after sunset on the day of his son's funeral. Grief was overtaking him.

If you give this emotion connected to foods development some thought, it begins to make sense. Consuming certain foods can produce personality traits or emotional outbursts, particularly foods in the category of animal products. Some examples were mentioned in the Introduction of this book but many more exist.

Banana	Stubbornness, frustration
Beef	Anger
Bread before sunset	Grief
Candies	Being upset
Chicken	Nervousness, anxious
Fruit breads and pastries	Griping, being grumpy
Lettuce	Bossy or grumpy
Marshmallow Cream	Grumpy
Peas	Jealousy
Sugar (cane and corn)	Anger
Vegetables	Verbal abuse

More importantly how does one go about erasing the collection of emotions?

I John 4:16-18: *And we have come to know and to believe the love that God has for us. God is love, and the one who remains in love remains in God and God remains in him. In this, love is perfected with us so that we may have confidence in the day of judgment, for we are as He is in the world. There is no fear in love; instead, perfect love drives out fear, because fear involves punishment. So the one who fears has not reached perfection in love.* (HCS)

If you have read *Living by the Light of the Moon* published by Harvest of Healing, LLC in 2024, you will have a better grasp on what the verses in I John are speaking of. Love is the act of exchange of Heavenly Gases between Heaven and a person. A process I compared to photosynthesis takes place when the Heavenly Gases are received, processed and released, returning to the environment. The Heavenly Gases are transported to our atmosphere from the cosmos, originating in the planets that are constructed of gases. When the Heavenly Gases are received, the person experiences a form of God assigned Heavenly Housekeeping, erasing all harmful signals or vibrations encountered. The love mentioned in I John is the Love Exchange that brings about the Heavenly Housekeeping. When the Heavenly Housekeeping is done properly and regularly, there will be no judgment (consequence) that comes against the flesh; any emotion of fear, or any other emotion, will not be present. Peace will reign. Punishment rides in the same category as judgment, both being a consequence to the physical body in the form of distress, sickness or disease. For additional information on emotions see, *From AntiChrist to I AM* published by Harvest of Healing, LLC in 2022.

THE LAST TRUMPET
I Corinthians 15:51-57:

Listen! I am telling you a mystery: We will not all fall asleep, but we will all be changed, in a moment, in the blink of an eye, at the last trumpet. For the trumpet will sound, and the dead will be raised incorruptible, and we will be changed. For this corruptible must be clothed with incorruptibility, and this mortal must be clothed with immortality. When this corruptible is clothed with incorruptibility, and this mortal is clothed with immortality, then the saying that is written will take place:

> Death has been swallowed up in victory.
> Death, where is your victory?
> Death, where is your sting?

Now the sting of death is sin, and the power of sin is the law. But thanks be to God, who gives us the victory through our Lord Jesus Christ! (HCS)

CONCLUSION

Dinner Bells

The meals, the feasts and tasty dishes,
All in response to human wishes.

Puddings, cakes, pies and more,
Who doesn't love them in all their galore.

Chocolate chip cookies and birthday cakes too,
No one knew what all it would do.

Put it all together and you'll see what's in store,
Not what you'd expect and this, I am sure.

Sit at a table and take it all in,
Who would have known it ends up as sin.

That sin takes a ride through the DNA chain,
To the next generation who wish you'd refrained.

Saddled with the burden of ancestor's eating,
who must have been charged with gluttony and cheating.

Paying their debt is nothing too easy,
But God has a plan that makes it sound breezy.

Clouds by day and a fire by night,
Erasing the sin that has caused such a fright.

The Spirit we need for a disease-free life,
Was taken from His People and resulted in strife.

The return of the Spirit is upon us now,
Be aware and alert and you'll see how.

Living Water is what we must have,
To capture and erase all that is bad.

The Light will become evident and begin to show,
The results of living according to what I've come to know.

This knowing is deep and comes from within,
The sin now erased so my Life can begin.

Izauh 61™

RESOURCES

Holy Bible: Holman Christian Standard; Christian Standard Bible; New King James Version; New Living Translation; New International Version.

Etymonline.com
NIH.gov
Gastro.org
Sciencenotes.org
The Golden Book of the Holy Vedas
The Laws of the Universe

Suggested Reading:

From AntiChrist to I AM
Food for the Journey to I AM
Published 2022, Harvest of Healing, LLC

Home-Made Answers for Cancer and Life Altering Disease
Published 2024, Harvest of Healing, LLC
Living by the Light of the Moon
Published 2024, Harvest of Healing, LLC

Milton Keynes UK
Ingram Content Group UK Ltd.
UKHW040734141124
451073UK00006BA/142